Green Finance and Investment

I0751829

Green Investment Banks

SCALING UP PRIVATE INVESTMENT IN LOW-CARBON, CLIMATE-RESILIENT INFRASTRUCTURE

This work is published under the responsibility of the Secretary-General of the OECD. The opinions expressed and arguments employed herein do not necessarily reflect the official views of OECD member countries.

This document and any map included herein are without prejudice to the status of or sovereignty over any territory, to the delimitation of international frontiers and boundaries and to the name of any territory, city or area.

Please cite this publication as:
OECD (2016), *Green Investment Banks: Scaling up Private Investment in Low-carbon, Climate-resilient Infrastructure*, Green Finance and Investment, OECD Publishing, Paris.
http://dx.doi.org/10.1787/9789264245129-en

ISBN 978-92-64-24511-2 (print)
ISBN 978-92-64-24512-9 (PDF)

Series: Green Finance and Investment
ISSN 2409-0336 (print)
ISSN 2409-0344 (online)

Photo credits: Cover © mika48/Shutterstock.com.

Corrigenda to OECD publications may be found on line at: *www.oecd.org/about/publishing/corrigenda.htm*.

© OECD 2016

You can copy, download or print OECD content for your own use, and you can include excerpts from OECD publications, databases and multimedia products in your own documents, presentations, blogs, websites and teaching materials, provided that suitable acknowledgement of OECD as source and copyright owner is given. All requests for public or commercial use and translation rights should be submitted to *rights@oecd.org*. Requests for permission to photocopy portions of this material for public or commercial use shall be addressed directly to the Copyright Clearance Center (CCC) at *info@copyright.com* or the Centre français d'exploitation du droit de copie (CFC) at *contact@cfcopies.com*.

Preface

Following the successful climate change agreement reached in Paris at the 21st Conference of the Parties to the United Nations Framework Convention on Climate Change (COP21), attention needs to shift quickly to how countries will achieve their Nationally Determined Contributions. Governments will need to take actions that will help accelerate a shift away from investments in carbon-intensive infrastructure and toward low-carbon, climate-resilient (LCR) infrastructure. Investment is growing in renewable energy and energy efficiency, but not quickly enough to get the world on track to achieve zero net greenhouse gas emissions globally by the end of this century. This "decarbonisation" of the global economy will be necessary if we are to hold the increase in the global average temperature to well below 2°C above preindustrial levels, as 195 countries agreed in Paris. To achieve these very ambitious goals, governments need to make full use of their capacity to leverage and unlock much larger flows of private investment in low-carbon infrastructure.

The OECD's work on green finance and investment aims to help governments overcome investment barriers, implement effective policies to drive low-carbon investment and innovation, understand and promote the development of investment channels, and consider the role of public interventions and institutions to mobilise private investment. This report focuses on a relatively new type of institution – publicly capitalised green investment banks (GIBs). Over a dozen national and sub-national governments have created GIBs and GIB-like entities, which are established specifically to facilitate private investment into domestic LCR infrastructure. Using innovative transaction structures, risk-reduction and transaction-enabling techniques, and local and market expertise, GIBs are channelling private investment into low-carbon projects. GIBs are facilitating investment in such areas as commercial and residential energy efficiency retrofits, rooftop solar photovoltaic systems and municipal-level, energy-efficient street lighting.

Although their common objective is to facilitate low-carbon investment, GIBs have been created in a variety of national and local contexts to achieve a range of goals, including meeting ambitious emissions targets, supporting local community development, lowering energy costs, developing green technology markets, creating jobs and lowering the cost of capital.

Green investment banks are not the only institutional option available to governments seeking to accelerate investment into domestic, low-carbon, climate-resilient infrastructure. Some national development banks have been providing financing for low-carbon projects for many years, as examined in previous OECD work on the role of public financial institutions in the low-carbon transition. Rather than offering prescriptions, this report offers a stock-taking on GIBs, their objectives, mandates, interventions and performance tracking.

Institutions like GIBs may best be understood as a tool to mobilise private investment which can complement climate policies but cannot substitute for them. Enabling policies for low-carbon investment – including a robust and credible carbon price, fossil fuel subsidy reform, well-designed renewable energy incentive policies and clear, long-term climate policy goals – are essential. But GIBs and other institutions can play a supportive role in overcoming remaining investment barriers. To mount a serious effort to mobilise low-carbon investment and get on a path toward zero net emissions by the end of this century, governments need to consider how institutions like green investment banks can help them pick up the pace.

Simon Upton
Director, OECD Environment Directorate

Foreword

This report aims to provide policy makers with the first comprehensive study of publicly capitalised green investment banks (GIBs), examining the rationales, mandates and financing activities of this relatively new category of public financial institution. It provides a non-prescriptive stock-taking of the diverse ways in which these public institutions are helping to leverage and catalyse private investment in domestic green infrastructure, with a spotlight on energy efficiency projects. Highlighting the role of GIBs within a broader policy framework to mobilise investment, the report also provides practical information to policy makers on how green investment banks are being set up, capitalised and staffed.

Developed by the Secretariat for the Working Party on Climate Investment and Development (WPCID) of the Environmental Policy Committee (EPOC), the report has linkages to previous and ongoing OECD work on improving policy frameworks for investment in low-carbon and climate-resilient (LCR) infrastructure and on the role of institutional investors in financing the low-carbon transition. For example, a 2015 OECD report entitled *Mapping Channels to Mobilise Institutional Investment in Sustainable Energy* recommended that governments consider the case for establishing a special-purpose, domestically focused green investment bank or refocusing the activities of existing public financial institutions to encourage greater institutional investment in green infrastructure. As several green investment banks focus on mobilising institutional investment, this report was also presented to the G20/OECD Task Force on Institutional Investors and Long-term Financing.

This report seeks to contribute to emerging literature on green investment banks and on the role of institutional investors in financing long-term investment more broadly. It also seeks to complement work focused on the actual and potential use of innovative financing instruments and risk mitigants to catalyse private investment in low-carbon, climate-resilient infrastructure.

GIBs are a relatively new type of institution, and as such, they have not yet been the focus of much analysis. For example, the literature analysing the performance of public financial institutions in crowding in (and avoiding crowding out) private investment has not yet addressed GIBs. As GIBs progressively develop a track record and more experience in leveraging investment in different technologies, future OECD research could assess the effectiveness of GIBs in cost-effectively mobilising private investment, avoiding crowding out private investment, carefully gauging investment risks, effectively targeting and addressing key investment barriers, and successfully demonstrating the viability of LCR infrastructure investment. Future research could also examine in greater detail the advantages and disadvantages of creating a green investment bank relative to mainstreaming green investment objectives in existing institutions such as national development banks.

Acknowledgements

This report was produced in the OECD Environment Directorate under the direction of Simon Upton and was completed as part of the OECD Environment Policy Committee's 2015-16 Programme of Work and Budget. The report was written by Kate Eklin and Robert Youngman with significant contributions from Jeffrey Schub (Coalition for Green Capital) and under the guidance of Simon Buckle (Head of the Climate, Biodiversity and Water Division of the Environment Directorate).

The joint project with Bloomberg Philanthropies was initiated by Lamia Kamal-Chaoui, Senior Advisor in the Office of the Secretary-General in charge of relationships with foundations, under the direct guidance of Gabriela Ramos, Chief of Staff and Sherpa. The OECD would like to thank Bloomberg Philanthropies for generous support for and substantive contributions to this project, with special thanks to Kevin Sheekey and Meridith Webster. A separate "Policy Perspectives" brochure on green investment banks, which distils key messages from this report, was produced jointly with Bloomberg Philanthropies and benefited from the input and advice of Daniel Firger, who also co-hosted a joint OECD-Bloomberg Philanthropies side event at the OECD Pavilion at the 21st Conference of the Parties to the United Nations Framework Convention on Climate Change (COP21). We would also like to thank Lee Cochran for her contributions to the OECD's outreach and communications for the project.

The report benefited from insights gained from the OECD's Green Investment Financing Forums (in June 2014 and May 2015) and Green Investment Bank Workshop (May 2015). It also draws on broader work undertaken by the OECD on "Public Policies for Facilitating Green Long-Term Infrastructure Investment", which is generously supported by voluntary contributions from the Japanese Ministry of Finance.

This report also benefited from review and comments provided by the OECD Environment Policy Committee (EPOC) and its Working Party on Climate, Investment and Development (WPCID), and the G20/OECD Task Force on Institutional Investors and Long-term Financing. The authors would like to thank their colleagues at the OECD and IEA Secretariats who provided valuable comments and review to the report: Geraldine Ang, Heymi Behar, Philippe Benoit, Simon Buckle, Jane Ellis, Raphaël Jachnik, Osamu Kawanishi, Tomasz Kozluk, Lorcan Lyon, Sara Moarif, Michael Mullan, Paul O'Brien, Chung-a Park, Mikaela Rambali, Dirk Röttgers, Michael Waldron and Simon Upton. The authors also would like to thank Christopher Kaminker for his research on green banks which provided an important foundation for this report, Bérénice Lasfargues for her editorial and substantive contributions to the publication, Jennifer Allain and Janine Treves for their valuable editorial guidance, and Jennifer Calder and Pascale Rossignol for their assistance with formatting the manuscript. In addition, the authors would like to thank the following expert reviewers for their inputs, comments and guidance to the report: Simon Brooker, Simon Every, Tristan Knowles, Meg McDonald, Marianna O'Gorman and Oliver Yates (Clean Energy Finance Corporation); Bryan Garcia and Bert Hunter (The

Connecticut Green Bank); Takejiro Sueyoshi (Green Finance Organisation, Japan); Syed Ahmad Syed Mustafa (GreenTech Malaysia); Tan Yan Chen, Mark Glick and Alan Yonan (Hawaii Department of Business, Economic Development & Tourism); Takuya Hosogai, Masanori Nakagawa and Yuichi Shiraishi (Japanese Ministry of Environment); Caroline Angoorly and Sarah Davidson (NY Green Bank); Silvia Ruprecht-Martignoli (Swiss Federal Department of the Environment, Transport, Energy and Communications); Jeremy Burke and Gavin Templeton (UK Green Investment Bank); and Joanne Lawson (UK Department for Business, Innovation & Skills).

Table of contents

Tables

Figures

Boxes

Follow OECD Publications on:

 http://twitter.com/OECD_Pubs

 http://www.facebook.com/OECDPublications

 http://www.linkedin.com/groups/OECD-Publications-4645871

 http://www.youtube.com/oecdilibrary

http://www.oecd.org/oecddirect/

Abbreviations and acronyms

AIIB	Asian Infrastructure Investment Bank
AUD	Australian dollar
B20	Business 20
BRICS	Brazil, Russia, India, China and South Africa
CCICED	China Council for International Cooperation on Environment and Development
CEFC	Clean Energy Finance Corporation (Australia)
CHP	Combined heat and power
CIF	Climate Investment Fund
COP	Conference of the Parties to the United Nations Framework Convention on Climate
CTF	Clean Technology Fund
EERS	Energy efficiency resource standards
ERB	Energy Resilience Bank (New Jersey, United States)
ESCO	Energy service company
EUR	Euro
G20	Group of Twenty
GCF	Green Climate Fund
GEF	Global Environment Facility
GEMS	Green Energy Market Securitization (Hawaii, United States)
GHG	Greenhouse gas
GIB	Green investment bank
GTFS	Green Technology Financing Scheme (Malaysia)
IDB	Inter-American Development Bank
IDFC	International Development Finance Club
IEA	International Energy Agency
LCR	Low-carbon and climate-resilient
LED	Light emitting diodes
MDB	Multilateral development bank

MW	Megawatt
NDB	National development bank
NYSERDA	New York State Energy Research and Development Authority
OECD	Organisation for Economic Co-operation and Development
PEV	Plug-in electric vehicle
PFI	Public financial institution
PPA	Power purchase agreement
PV	Photovoltaic
R&D	Research and development
RGGI	Regional Greenhouse Gas Initiative (New York, United States)
RPS	Renewable portfolio standards
SCCF	Special Climate Change Fund
UKWREI	UK Waste Resources and Energy Investments
UN	United Nations
UNFCCC	United Nations Framework Convention on Climate Change
USD	United States dollar
WHEEL	Warehouse for Energy Efficiency Loans (United States)

Executive summary

Despite growing investment in renewable energy and energy efficiency, efforts to significantly scale up private investment in green infrastructure, including low-carbon and climate-resilient (LCR) infrastructure, continue to face challenges. Pricing signals often favour investment in unabated fossil-fuel intensive activities over LCR alternatives since the social costs of emissions are not adequately reflected and even commercially viable LCR projects can be associated with higher risks and transaction costs. As governments work to meet their pre- and post-2020 emission reduction pledges, they will need to make efficient use of public funding to mobilise much larger amounts of private investment in LCR infrastructure.

To overcome investment barriers and leverage the impact of available public resources, over a dozen national and sub-national governments have created public green investment banks (GIBs) and GIB-like entities in recent years. A GIB is a publicly capitalised entity established specifically to facilitate private investment into domestic LCR infrastructure and other green sectors such as water and waste management. These dedicated green investment entities have been established at national level (Australia, Japan, Malaysia, Switzerland, United Kingdom), state level (California, Connecticut, Hawaii, New Jersey, New York and Rhode Island in the United States), county level (Montgomery County, Maryland, United States) and city level (Masdar, United Arab Emirates).

While GIBs differ in name, scope and approach, they generally share the following core characteristics: a mandate focusing mainly on mobilising private LCR investment using interventions to mitigate risks and enable transactions; innovative transaction structures and market expertise; independent authority and a degree of latitude to design and implement interventions; and a focus on cost-effectiveness and performance. "GIB-like entities" refers to organisations that have a mandate to leverage private finance for domestic LCR infrastructure investment but which may not possess all of the core characteristics of GIBs and may pursue other activities or use other approaches.

Based on their unique national and local contexts, governments tailor their GIBs. GIBs and GIB-like entities have diverse rationales and goals, including meeting ambitious emissions targets, mobilising private capital, lowering the cost of capital, lowering energy costs, developing green technology markets, supporting local community development and creating jobs. These goals are reflected in the range of metrics GIBs use to measure and track their performance and demonstrate accountability: emissions saved, job creation, leverage ratios (i.e. private investment mobilised per unit of GIB public spending) and, in some cases, rate of return.

Governments are using GIBs to channel private investment, including from institutional investors, into low-carbon projects such as commercial and residential energy efficiency retrofits, large-scale onshore and offshore wind, rooftop solar photovoltaic systems and municipal-level, energy-efficient street lighting. Unlike grant-making public institutions, GIBs focus on financial sustainability and some are required to be profitable. For example, the UK Green Investment Bank must invest on commercial terms and has to meet a minimum 3.5% annual nominal return on total investments, after operating costs but before tax. Through their interventions and investments, GIBs are demonstrating to private investors that commercially successful investments are possible and happening now.

Governments have capitalised GIBs using a variety of funding sources including: government appropriations and programmes (including reallocation of funds from existing programmes); revenue from carbon taxes, emissions trading schemes, renewable portfolio standards and energy efficiency resource standards; utility bill charges; and bond issuance. GIBs are typically smaller than national development banks and other public financial institutions that mobilise private investment in domestic LCR infrastructure. The size of the (currently) largest GIB is expected to be approximately USD 7 billion at final capitalisation, while Germany's KfW invested approximately USD 56 billion in 2015 in "domestic promotion", including but not limited to "special programmes to foster the use of renewable energy, to increase energy efficiency and to promote innovative technology companies". This smaller size is not preventing GIBs from mobilising significant private investment, however. Some GIBs like the UK Green Investment Bank, Australia's Clean Energy Finance Corporation and the Connecticut Green Bank are successfully targeting institutional investors – notably pension funds, insurance companies, sovereign wealth funds and mutual funds – for co-investment in funds and other transactions. These investors represent a large pool of capital and an increasingly important alternative source of financing for LCR infrastructure investment, as examined in other OECD reports.

This report also draws particular attention to the role of GIBs in attracting private investment in energy efficiency. This is relevant to the OECD's ongoing work on energy efficiency financing, including support to the G20 Energy Sustainability Working Group (ESWG). GIBs are designed to address a range of investment barriers to energy efficiency through a variety of interventions, such as creating funds; providing direct corporate loans, leasing and loan warehousing; and offering on-bill finance, where borrowers can repay an energy efficiency project through savings on their existing utility bills. Another approach is to link energy efficiency loan repayment to property tax payments through tax liens (e.g. "Property-Assessed Clean Energy" (PACE) in the United States). This approach facilitates investment by allowing energy savings to offset loan repayments, while making repayment effortless for borrowers and creating increased security for lenders. For example, the Connecticut Green Bank's C-PACE programme financed, in less than two years, nearly USD 54 million in energy upgrades for 89 buildings, accounting for about one-third of the commercial PACE market in the United States.

GIBs are a tool to mobilise private investment that can complement but not replace climate policies such as putting a price on carbon and reforming inefficient fossil-fuel subsidies. Well-designed climate policies create many of the conditions necessary to stimulate LCR investments. Within such a framework, GIBs can play a supportive role in overcoming remaining barriers and catalysing investment. In addition to GIBs, other institutional options are available to governments seeking to catalyse green investment, such as mainstreaming green investment in existing national development banks. Nevertheless, GIBs are making a case that centralising expertise in a new independent institution dedicated to mobilising green private investment can be an effective approach to unlocking larger flows of private capital.

Chapter 1.

Using green investment banks to scale up private investment

This chapter introduces green investment banks as a relatively new type of institution focused on increasing private investment in domestic low-carbon and climate-resilient infrastructure and other green sectors. Given the variety of existing public and private financial institutions that support green infrastructure investments, the chapter situates green investment banks within this wider context. The chapter closes with a discussion of factors governments may consider when evaluating the need to create a green investment bank. The chapter serves as a detailed introduction to green investment banks for policy makers and as an extended summary of the main messages of the report.

Introducing green investment banks

Given the urgent need to accelerate the transition to a low-carbon economy, governments are increasingly focused on finding ways to leverage greater amounts of private investment in domestic low-carbon and climate-resilient (LCR) infrastructure (Box 1.1). In recent years an increasing number of governments have created green investment banks (GIBs) and GIB-like entities to help meet this objective. A GIB is defined for the purposes of this report as a publicly capitalised entity established specifically to facilitate and attract private investment into domestic LCR infrastructure and other green sectors such as water and waste management through different activities and interventions.

Box 1.1. Defining low-carbon and climate-resilient infrastructure investments and green infrastructure

Choices of infrastructure or selected features of infrastructure will affect the greenhouse gas emissions intensity of service provision (e.g. water, electricity, mobility, shelter, sanitation services) as well as the exposure and vulnerability of businesses and people to climate change itself. Low-carbon and climate-resilient (LCR) infrastructure projects either mitigate greenhouse gas emissions or support adaptation to climate change or both.

In addition to renewable energy, the term green infrastructure can cover a broad range of investments, including sustainable agriculture, floodplain levees and coastal protection, waste management infrastructure and "green" water infrastructure. Green water infrastructure may include wastewater treatment and infrastructure that requires less concrete, e.g. through rainwater harvesting, source control of surface water (such as sustainable urban drainage systems), green roofs, and local processing of grey or black water.

This report focuses mainly on a subset of green infrastructure investments, namely LCR investments made in companies, projects and financial instruments that operate primarily in the renewable energy, renewable technology and environmental technology markets as well as those investments that are climate-change specific. These investments include energy efficiency projects, many types of renewable energy generation, carbon capture and storage, smart grids and electricity demand-side management technology, and new transport technologies (e.g. electric vehicles).

Sources: Corfee-Morlot, J. et al. (2012), "Towards a green investment policy framework: The case of low-carbon, climate-resilient infrastructure", *OECD Environment Working Papers*, No. 48, OECD Publishing, Paris, http://dx.doi.org/10.1787/5k8zth7s6s6d-en; Kennedy, C. and J. Corfee-Morlot (2012), "Mobilising investment in low carbon, climate resilient infrastructure", *OECD Environment Working Papers*, No. 46, OECD Publishing, Paris, http://dx.doi.org/10.1787/5k8zm3gxxmnq-en.

A key factor informing decisions to create green investment banks is the presence of barriers to investment in LCR infrastructure. Some of these barriers are broadly applicable to low-carbon investments, such as: a failure to sufficiently price fossil-fuel externalities or to reform inefficient fossil-fuel support measures; a lack of suitable financial instruments with attributes sought by private investors; and a shortage of objective information, data and skills to assess transactions and underlying risks, among others (Box 1.2). Other barriers are specific to energy efficiency investment, including: small average investment size, relatively high transaction costs and the corresponding need to aggregate projects; the need to structure investments for retail and commercial energy efficiency to allow energy savings to offset loan repayments; and the tendency for local lenders to focus only on a borrower's credit rating during the underwriting process for an energy efficiency loan, rather than the project's estimated energy savings.

Box 1.2. Barriers to scaling up low-carbon and climate-resilient infrastructure investment

A range of barriers can affect the risk-return profile of low-carbon and climate-resilient (LCR) infrastructure and can determine whether LCR infrastructure investments are attractive or accessible to investors.

Barriers to scaling up LCR infrastructure include, but are not limited to, the following:

1. Environmental, energy and climate policies and regulations that favour investment in unabated fossil-fuel intensive activities over green infrastructure

Inconsistent policy signals can limit the attractiveness of green infrastructure for investors. These include continuing support for fossil-fuel use and production, low or no prices on greenhouse gas (GHG) emissions and unpredictable changes to support policies for renewable energy generation.

2. Regulatory policies with unintended consequences

The global financial crisis has motivated changes to financial stability rules and prudential regulation (e.g. Basel III and Solvency II) that may inadvertently limit the ability of regulated institutions such as banks and insurance companies to finance long-term infrastructure investments. Financial stability is a prerequisite to any kind of investment, and to this end, strengthening the resilience of banks through higher capital and liquidity requirements as well as structural reforms, and more monitoring of system-level risks by financial supervisors are critical. At the same time, a review and evaluation of the impacts of regulations on long-term finance is important to spot and evaluate potential consequences for the supply of long-term finance that will be needed for low-carbon investment.

3. A lack of suitable financial instruments and funds with attributes sought by private investors

Few LCR infrastructure financial instruments and funds have the necessary attributes of familiarity, investment-grade credit rating, low transaction costs, liquidity, appropriate investment period and availability of related financial research that will make them attractive to private investors.

4. A shortage of objective information, data and skills to assess transactions and underlying risks

In the absence of transparent information, data and financial research about LCR infrastructure that can act as a signal to investors or means for performance comparison in any given sector, there are significant barriers to entry. Unlike such investments as stocks, bonds and real estate investment trusts, green infrastructure and infrastructure investment performance data are generally not collected systematically.

The OECD report *Mapping Channels to Mobilise Institutional Investment in Sustainable Energy* highlights the barriers that specifically limit institutional investor investment in sustainable energy projects (OECD, 2015a).

Sources: OECD (2013a), "Long-term investors and green infrastructure: Green infrastructure", Policy Highlights Brochure, OECD, Paris; OECD (2015a), *Mapping Channels to Mobilise Institutional Investment in Sustainable Energy*, Green Finance and Investment, OECD Publishing, Paris, http://dx.doi.org/10.1787/9789264224582-en.

To address some of these barriers, GIBs employ a variety of techniques ("risk mitigants") that aim to mitigate risk and enable a larger flow of deals than would otherwise occur. More specifically, they use a range of targeted interventions to reduce, reassign or reapportion different investment risks using mechanisms such as guarantees, insurance products, public stakes and other forms of credit enhancement. By providing coverage for risks which are new and are not currently covered by financial actors, or are simply too costly for investors, risk-mitigating tools increase the attractiveness and acceptability of investments (OECD, 2015a).

Other GIB techniques seek to reduce transaction costs. As many investors have limited experience with investment in LCR infrastructure, the cost associated with identifying, executing and managing such investments can be prohibitive. In addition, LCR infrastructure investments – and particularly energy efficiency investments – are typically too small to be attractive to many private investors due to high transaction costs. To reduce these costs, GIBs employ various approaches ("transaction enablers"), including warehousing (pooling small transactions), securitisation (transforming illiquid assets into tradable securities) in a prudent and judicious way and co-investment (OECD, 2015a). The OECD report *Mapping Channels to Mobilise Institutional Investment in Sustainable Energy* explores risk mitigants and transaction enablers in detail (OECD, 2015a).

In addition to using these techniques, GIBs seek to prove through "demonstration" that LCR infrastructure investments can be profitable today on commercial terms, even without risk mitigation. Demonstration aims to: address incorrect perceptions among investors that clean technologies are less developed, risky and not commercially viable; fill data and information gaps; and build confidence in markets for new technologies and activities.

Table 1.1 lists the GIBs and "GIB-like entities" discussed in this report.[1] "GIB-like entities" refers to organisations that have a mandate to leverage private finance for domestic LCR infrastructure investment, but which may not possess all core characteristics of GIBs, and may pursue other activities or use other approaches (e.g. grants).

Table 1.1. **Green investment banks or green investment bank-like entities in operation**

Operational green investment banks (GIBs) and GIB-like entities	Location	Year of formation
California CLEEN Center	California, United States	2014
Clean Energy Finance Corporation (CEFC)	Australia	2012
Connecticut Green Bank	Connecticut, United States	2011
Green Energy Market Securitization (GEMS) (Hawaii Green Infrastructure Authority)	Hawaii, United States	2014
Green Fund	Japan	2013
Malaysian Green Technology Corporation (GreenTech Malaysia)	Malaysia	2010
Masdar	United Arab Emirates	2006
New Jersey Energy Resilience Bank (ERB)	New Jersey, United States	2014
NY Green Bank	New York, United States	2014
Rhode Island Infrastructure Bank (RIIB)	Rhode Island, United States	2015
Technology Fund	Switzerland	2014
UK Green Investment Bank	United Kingdom	2012

Individual governments' rationales and motivations for creating GIBs vary. In addition to reducing greenhouse gas (GHG) emissions, policy makers have cited factors such as local and regional development, global competitiveness, energy security and job creation as important reasons for establishing a GIB (Table 1.2). Despite these varying rationales, GIBs share an underlying goal – to address investment barriers and catalyse private investment in LCR infrastructure.

The UK government created the UK Green Investment Bank in 2012 as a tool to develop markets and cost effectively meet its legally binding GHG reduction targets established in 2008 (Green Investment Bank Commission, 2010). Australia's Clean Energy Finance Corporation (CEFC) was created in 2012 under similar circumstances, as part of a national climate policy scheme that at the time included a carbon pricing plan. Malaysia's Green Technology Financing Scheme was established to increase the development and use of green technology (OECD, 2013b) and was created as part of the broader GreenTech Malaysia organisation, which has a multi-pronged mission to promote environmental, economic and social well-being. The Japanese Green Fund was created to reduce carbon dioxide (CO_2) emissions.

Table 1.2. **Summary of rationales for creating green investment banks (GIB) and GIB-like entities**

Entity	Capital market efficiency	Reduce greenhouse gas emissions	Lower price of energy	Increase grid reliability	Create jobs/ industry growth	Part of national climate policy	Increase sustainability
California CLEEN Center (California, United States)		X			X	X	X
Clean Energy Finance Corporation (CEFC) (Australia)	X	X		X		X	
Connecticut Green Bank (Connecticut, United States)	X	X	X	X	X		
Green Energy Market Securitization (GEMS) (Hawaii Green Infrastructure Authority) (Hawaii, United States)	X		X				
Green Fund (Japan)		X			X	X	
Malaysian Green Technology Corporation (GreenTech Malaysia) (Malaysia)		X			X	X	X
Masdar (United Arab Emirates)	X				X		X
New Jersey Energy Resilience Bank (New Jersey, United States)				X			
NY Green Bank (New York, United States)	X	X		X			
Technology Fund (Switzerland)		X			X	X	
UK Green Investment Bank (United Kingdom)	X	X			X	X	X

In New York, the state government established NY Green Bank in 2013 because it wanted public funding that had previously been used almost exclusively for grant programmes to go further and attract greater private investment. The Connecticut Green Bank's goal when it was created in 2011 was to make power "cheaper, cleaner and more reliable" for a state which then had the third highest electricity costs in the United States (US EIA, 2012). Responding to the need to make private rooftop solar photovoltaic (PV) more accessible, Hawaii's Green Energy Market Securitization programme (GEMS) was designed in 2013 to increase the availability of financing, particularly for underserved markets, including renters, low-income individuals, non-profit organisations and people not otherwise able to acquire renewable energy systems. GEMS was also established to help the state reach its renewable energy portfolio standard goals. While GEMS funding initially will be available for solar PV systems, it can also be used to finance energy storage, energy efficiency and other renewable energy technologies.

In New Jersey, energy security and the development of climate-resilient energy infrastructure were central to the decision in 2014 to create the New Jersey Energy Resilience Bank (ERB). This bank was established to facilitate investment in and provide technical support to power platforms and critical infrastructure that could withstand high-impact weather events such as Hurricane Sandy (New Jersey Board of Public Utilities, 2014).

While GIBs differ in name, scope and approach, they generally share the following core characteristics:

- Narrow mandate: GIBs generally have a narrow mandate focusing mainly on mobilising private LCR investment (but sometimes on broader green infrastructure investment) using interventions to mitigate risks and enable transactions.
- Independence: GIBs are typically established as special-purpose public or quasi-public entities which are granted independent authority to meet their mandates and a degree of latitude to design and implement interventions based on their deal-making and sectoral expertise.
- Additionality: GIBs seek to provide additional capital to facilitate transactions that would not occur without GIB involvement.
- Cost-effectiveness: GIBs mobilise private capital using least-cost solutions in order to reduce public expenses or as part of an organisational mandate for profitability.
- Accountability: GIBs are evaluated using metrics such as the amount of private capital mobilised, return on capital, number of jobs created and GHG reductions. GIBs' public reporting on their performance typically includes transparent calculation methodologies to build credibility.

Some other characteristics of green investment banks are summarised in Box 1.3. Given that GIBs' track record is still relatively limited and this report is principally a stock-taking rather than an assessment, further research would be needed to evaluate the performance of GIBs. For example, future work could usefully focus on gathering and evaluating evidence of GIBs' performance with respect to cost-effectiveness, avoiding crowding out private investment, carefully gauging investment risks, effectively targeting and addressing key investment barriers, and successfully demonstrating the viability of LCR infrastructure investment.

In addition to GIBs that possess all of the core GIB characteristics, there are other domestic institutions that could be described as "GIB-like entities". These institutions incorporate some elements of GIBs but differ in other areas. For example, renewable energy funds or programmes, such as the Swiss Technology Fund, may use some of the same interventions to mobilise private investment that GIBs use, but do not have the independence of GIBs to select and structure different interventions. The Connecticut Green Bank was initiated as a renewable energy fund and was converted into a GIB in order to expand its activities and provide a more rigorous mandate and greater independence so as to take advantage of its experience and expertise. Another example of a GIB-like entity is Masdar (Abu Dhabi, United Arab Emirates), which has some subsidiaries which carry out GIB-like activities while others use a different model. For instance, Masdar Clean Energy, similar to a GIB, focuses on investments in renewable energy projects using commercial technologies. Masdar Capital, on the other hand, operates more like a private equity fund.[2] Like many GIBs, it has an objective of profitability, but it has more capacity to take risks (e.g. those associated with earlier-stage technologies) than a GIB that must meet specific requirements for financial performance.

Box 1.3. **How green investment banks view their added value**

Overcoming investment barriers: Green investment banks (GIBs) typically have a specific mandate to overcome barriers to scaling up low-carbon and climate-resilient (LCR) infrastructure investment. They use targeted approaches and tailored financial structuring to address the lack of suitable LCR investments with attributes sought by private investors (e.g. through aggregation of small-scale investments such as residential rooftop solar photovoltaic [PV] investments or energy efficiency retrofits in commercial buildings). They also address a shortage of objective information, data and skills to assess transactions and underlying risks. GIBs work with market participants to increase the supply of and demand for profitable low-carbon investments by decreasing risks, increasing market transparency and improving investors' (including lenders') understanding of low-carbon investments.

Building confidence by reducing risk: Mainstream lenders and investors can be slow to gain confidence in new technologies. GIBs accelerate the process by reducing real and perceived risk and increasing the number of transactions in markets for new technologies.

Relying on local expertise: GIBs hire financial professionals with local and national expertise in low-carbon technologies, projects and investments, and an understanding of the specific risk-return appetites of local financial institutions and other investors such as institutional investors. This local expertise provides informational advantages that can be leveraged to overcome investment barriers, which are often location-specific.

Transforming markets: GIBs typically aim to demonstrate the profitability of low-carbon investments to accelerate market development and then move on to other investments where they can improve the risk-return profile and attract private investment. GIBs are better placed to play this role than traditional government programmes – which may be less flexible and less familiar with markets – and than private companies – which face competitive pressures.

Reducing local financing costs: By dispersing information, sharing expertise and demonstrating that investments are profitable, GIBs help accelerate reductions in financing costs.

Source: Personal communication with Douglass Sims, Natural Resource Defense Council, October 2015; OECD Green Investment Bank Workshop, 22 May 2015.

The People's Republic of China (hereafter "China") is a prominent example of a country with market and institutional settings that are distinct from those in most other countries with GIBs, and which is considering the establishment of a GIB or GIB-like entity (Box 1.4).

Investments and policies needed to meet a 2°C target

Climate objectives and LCR investment needs and challenges form the backdrop for governments' interest in creating GIBs and supporting other efforts to mobilise private investment in LCR infrastructure. In the Paris Agreement adopted in December 2015 by the 21st Conference of the Parties to the United Nations Framework Convention on Climate (COP21), parties agreed to transition to "aggregate emission pathways consistent with holding the increase in the global average temperature to well below 2°C above preindustrial levels" (UNFCCC, 2015). An estimated USD 93 trillion in infrastructure investments across transport, energy, water systems and cities will be needed over the period 2015 to 2030 to meet global infrastructure needs while ensuring the transition to a low-carbon economy (Global Commission on the Economy and Climate, 2014). Given that traditional sources of green infrastructure finance and investment – governments, commercial banks and utilities – face significant constraints, alternative sources will be needed not only to compensate for these constraints, but also to ramp up green infrastructure investments.[3] Due to the numerous barriers to scaling up LCR investment (Box 1.2), public interventions are needed to mobilise additional private investment in LCR infrastructure.

Box 1.4. A national green bank in China?

The China Council for International Cooperation on Environment and Development (CCICED) has recommended the creation of a National Green Development Fund. If implemented as proposed, the fund would have a capitalisation target of approximately RMB 300 billion (USD 47 billion) and could raise more private capital as required. The proposed fund would focus on providing equity investments to facilitate access to other financing, including bank loans. It would operate on a commercially sustainable basis and seek to pool capital from investors with differing risk and return requirements. Sources of capital for the fund could include "fiscal funds from the central government, development finance, and other interested financial institutions and private investors." Its focus would be on investments in "resource efficiency, renewable energy, industrial pollution control and advanced vehicle technologies" (CCICED, 2015).

Renewable energy investment needs in China are significant (USD 1 trillion of cumulative investment in wind and solar PV from 2014-35) (IEA, 2014). Investments could be accelerated by a national green bank and broader policies for green finance reform and green transformation recommended by the CCICED, including policies to develop the domestic green bond market.

Source: CCICED (2015), "Green financial reform and green transformation", report to the Annual Conference of CCICED, 9-11 November, China Council for International Cooperation on Environment and Development, www.cciced.net/encciced/policyresearch/report/201511/P020151117574533056430.pdf; IEA (2014), *World Energy Investment Outlook*, OECD/IEA, Paris, www.iea.org/publications/freepublications/publication/WEIO2014.pdf.

GIBs are a tool to mobilise private investment that can complement policies but cannot replace core climate policies. The OECD has developed policy frameworks and guidance which seek to integrate considerations of climate and investment policy in order to establish strong enabling conditions for investment (Box 1.5). To enable LCR investment, governments must send a robust and credible price signal to internalise the cost of greenhouse gas emissions, remove fossil fuel subsidies, provide incentives for renewable energy generation and set clear, long-term policy goals. When some or all of these conditions are in place, GIBs can play a supportive role in overcoming remaining barriers and catalysing investment. Policy makers considering a GIB should consider how the institution can be integrated with existing public policies and investment promotion initiatives.

Box 1.5. Green investment policy framework and policy guidance in renewable energy

The OECD has developed a five-point "green investment policy framework" that aims to integrate climate and investment policy to provide coherent incentives and establish strong enabling conditions for green investment in the domestic context:

1. Set clear, long-term strategic policy goals in infrastructure planning and climate policies.
2. Implement policies and incentives to support LCR investment, for instance by putting a price on carbon, and removing fossil fuel subsidies and providing well-designed, well-timed, well-targeted and time-limited incentives for renewable energy investment.
3. Provide the right financial instruments to reduce risk and increase returns of green infrastructure projects.
4. Harness resources (for instance in research and development) and build capacity.
5. Promote greener consumer and business behaviour.

Together, these elements of a green investment policy framework can help to mobilise private investment and bring transformational change.

Although domestic policies to promote renewable energy infrastructure have greatly expanded throughout the world, policies in a number of related areas can create significant barriers to the effectiveness of these policies and potentially to the efforts of a green investment bank. To identify and address potential roadblocks to mobilising private investment in renewable energy infrastructure in emerging and developing economies, the *OECD's Policy Guidance for Investment in Clean Energy Infrastructure* (OECD, 2015b) raises issues for policy makers' consideration in the areas of investment policy, investment promotion and facilitation, competition, financial market and public governance policies. Similarly, investments in sustainable transport infrastructure also have their own particular set of challenges and channels that a new GIB may need to take into account (see Ang and Marchal, 2013). Other policy misalignments in the electricity sector and the broader economy which impede the transition to a low-carbon economy are examined in greater detail in *Aligning Policies for a Low-carbon Economy* (OECD, 2015c).

Sources: Ang, G. and V. Marchal (2013), "Mobilising private investment in sustainable transport: The case of land-based passenger transport infrastructure", *OECD Environment Working Papers*, No. 56, OECD Publishing, Paris, http://doi.org/10.1787/5k46hjm8jpmv-en; Corfee-Morlot, J. et al. (2012), "Towards a green investment policy framework: The case of low-carbon, climate-resilient infrastructure", *OECD Environment Working Papers*, No. 48, OECD Publishing, Paris, http://dx.doi.org/10.1787/5k8zth7s6s6d-en; OECD (2015b), *Policy Guidance for Investment in Clean Energy Infrastructure: Expanding Access to Clean Energy for Green Growth and Development*, OECD Publishing, Paris, http://dx.doi.org/10.1787/9789264212664-en; OECD (2015c), *Aligning Policies for a Low-carbon Economy*, OECD Publishing, Paris, http://dx.doi.org/10.1787/9789264233294-en.

Situating green investment banks among other institutions mobilising private climate finance and investment

GIBs and GIB-like entities are situated within a broad spectrum of public institutions and entities that provide financing or leverage private climate finance and investment. They are mobilising private investment within a broader ecosystem of multilateral development banks (MDBs), national development banks (NDBs), bilateral development finance institutions, international climate funds and various private sector investors and financiers and investors (Annex 1.A1). To provide more context, Figure 1.1 illustrates a number of the diverse actors involved in LCR infrastructure investment financing. It also considers their respective focus on domestic vs. international investment and on "pure play"[4] LCR investment (i.e. an exclusive focus on LCR investment) vs. diversified infrastructure investment.

Figure 1.1. **Green investment banks and their relation to other existing public and private entities that finance low-carbon and climate-resilient infrastructure**

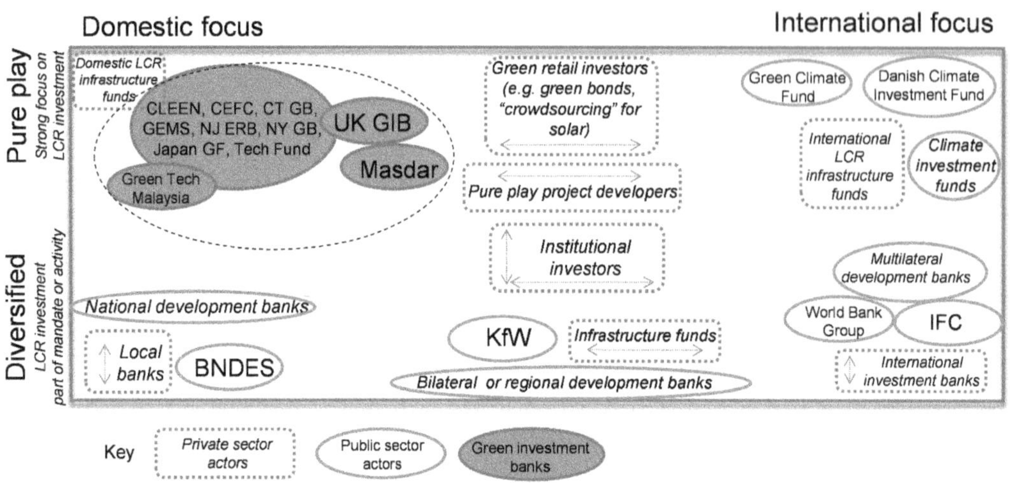

Note: BNDES: Brazilian Development Bank; CEFC: Clean Energy Finance Corporation; CLEEN: California CLEEN Center; CT GB: Connecticut Green Bank; GEMS: Green Energy Market Securitization; IFC: International Finance Corporation; Japan GF: Japan's Green Fund; LCR: low-carbon and climate-resilient; NY GB: NY Green Bank; Tech Fund: Technology Fund; NJ ERB: New Jersey Energy Resilience Bank; UK GIB: UK Green Investment Bank.

This figure is not intended to be exhaustive, but rather illustrative of certain institutions. The proximity of actors is not intended to reflect any particular interaction or co-operation but simply demonstrates the similar domains where public and private entities may pursue low-carbon and climate-resilient infrastructure investment. Arrows are used to illustrate that actors can be situated in different locations along each axis.

Source: Eklin, K. et al. (2016, forthcoming), "OECD Green Investment Financing Forum: Lessons from established and emerging green investment bank models", Background Note, OECD, Paris, forthcoming.

As shown in Figure 1.1, GIBs and GIB-like entities occupy the upper-left quadrant, which reflects their orientation (with exceptions) toward increasing private investment in domestic, pure-play LCR infrastructure. All GIBs and GIB-like entities focus on domestic infrastructure, with the exception of the pilot joint venture announced in March 2015 by the UK Green Investment Bank and the UK Department of Energy and Climate Change (DECC) to invest in India and Africa[5]. Many but not all GIBs and GIB-like entities focus exclusively on LCR infrastructure. Other sectors covered by GIBs and GIB-like entities include waste management (Japan's Green Fund), waste recycling and bioenergy (UK Green Investment Bank), and environmental mitigation and water treatment (California CLEEN Center).

In a short period of time (generally since 2011), GIBs and GIB-like entities have rapidly emerged as a new type of institution focusing on mobilising private investment. For policy makers, these entities merit consideration in light of their ability to be replicated in and adapted to different countries, at the national and sub-national level, and with a range of objectives.

The focus of green investment banks on domestic LCR infrastructure is another important feature. Although efforts to scale up private investment flows in LCR infrastructure need to focus on both domestic and international investment, flows of climate finance have been predominantly domestic to date. Total domestic climate finance flows – public and private flows combined – are more than double the size of cross-border flows (CPI, 2013; Hašcic et al., 2015). Private climate finance in particular is strongly oriented toward domestic investment. Ninety percent of private climate finance investments remained in their country of origin (CPI, 2014). GIBs' focus on and understanding of local markets and investment barriers are particularly relevant in this context. Governments will increasingly need to make efficient use of public funding to mobilise much larger amounts of private investment in their domestic LCR infrastructure; this is an important part of their broader effort to provide enabling policies for domestic low-carbon investment. For these and other reasons, this study aims to provide a stock-taking on GIBs to inform governments' further consideration of these entities as a potential tool to help meet emission reduction, investment mobilisation and other objectives.

At the same time, GIBs should be understood as being a new player in a broader ecosystem of generally much larger institutions and funds that are active in mobilising private LCR infrastructure investment. Annex 1.A1 describes these entities, which include: government-sponsored loan programmes; green programmes or initiatives within existing national development banks, bilateral development finance institutions, export credit agencies, multilateral development banks or central banks; multilateral infrastructure development banks and other infrastructure-specific initiatives, including the New Development Bank, the Asian Infrastructure Investment Bank and the Global Infrastructure Hub; agencies and institutions supporting research and development (R&D) and early development of clean technology; public agencies that implement national energy plans; and purely international climate funds.

Greening existing institutions versus establishing new ones

To mobilise private investment in domestic green infrastructure, "greening" existing institutions may be preferable to creating new institutions when the necessary institutional and political support exists. For example, many countries have NDBs (or public investment, infrastructure or industrial development banks) which focus on domestic investment. These banks are typically much larger than even the largest GIB. While many NDBs are less focused on mobilising green investment than GIBs, some NDBs have been providing financing for low-carbon projects for many years. For example, Germany's KfW has been investing in environmental protection domestically and internationally since the 1980s, and invested approximately USD 56 billion in 2015 in "domestic promotion", including but not limited to "special programmes to foster the use of renewable energy, to increase energy efficiency and to promote innovative technology companies" (KfW, 2016). Given the resources and longer track records of some NDBs in leveraging private climate finance and investment, they can provide important lessons for GIBs.

GIBs may also not be suitable for all countries. Establishing a GIB presumes a domestic context in which relatively limited interventions are sufficient to facilitate domestic private investment. Some domestic policy environments and local markets may be insufficiently developed to be appropriate for a GIB which uses commercial interventions. In these cases, market development and capacity building, and therefore grant models and significant subsidisation (e.g. from MDBs), are often required. However, the global spread of renewable energy markets may make GIBs (or GIB-like entities) potentially relevant for a large number of countries. One study estimates that "[a]s of early 2015, at least 164 countries had renewable energy targets, and an estimated 145 countries had renewable energy support policies in place" (REN21, 2015).

Some factors to consider when evaluating the relative benefits of creating a GIB or greening existing institutions include:

- **Costs:** Establishing a new institution likely involves more time and costs than greening an existing institution, and may be viewed as expanding bureaucracy or creating duplicative government services.
- **Independence and authority:** Creating a new GIB with an independent status can provide flexibility to experiment, innovate and adapt to market developments. It can also shield the institution from day-to-day political interference. In the case of the UK Green Investment Bank, this was deemed essential to attract long-term capital from institutional investors (UK House of Commons, 2011). Institutional barriers and political context could make it difficult for GIBs to address certain issues (CPI, 2015). Those barriers could apply equally to NDBs, however.
- **Mandate and culture:** Many NDBs lack a clear mandate to promote national climate change mitigation (Smallridge et al., 2013). NDBs may support renewable energy projects while also financing fossil fuel projects in parallel. In contrast, GIBs are exclusively focused on green investment and face fewer competing agendas.
- **Financing approaches:** The types of preferred financing approaches vary across GIBs, NDBs and MDBs. The International Development Finance Club (IDFC), which brings together over 20 NDBs and sub-regional development banks from around the world, estimates that members made new commitments representing USD 98 billion in green finance in 2014. Among the IDFC's members, 51% of financing in 2013 was in the form of non-concessional loans, followed by concessional loans (44%) and grants (3%). Other financial instruments such as equity, guarantees and unspecified loans accounted for only 2% of investment (IDFC, 2015). GIBs tend to be more oriented toward accelerating risk-taking by investors, through demonstration, co-investment and sharing risks with investors using guarantees and other risk mitigants. However, there are exceptions to these characterisations of NDBs and GIBs. Some NDBs, such as KfW, as well as multilateral development banks like the European Investment Bank and others, also increasingly develop and use innovative tools to scale up private finance from multiple investor classes. Some GIB-like entities (e.g. GreenTech Malaysia) make extensive use of concessional loans while GIBs like Australia's CEFC and Connecticut Green Bank use them only on a limited, targeted basis.
- **Scale:** The low-carbon investment portfolios of some NDBs are larger than those of even the largest GIB. If NDBs mainstream green investment throughout their portfolios, they may be able to mobilise LCR infrastructure at much greater scale than GIBs. However, if GIBs were able to significantly augment their current capitalisation by securing funds from other sources (e.g. the Green Climate Fund), the scale advantage held by NDBs could diminish.
- **Benefits of centralising green bank functions in one institution:** In addition to "greening" a single institution such as an NDB, another alternative to creating a GIB is to strengthen and expand green investment programmes that are already housed in different government agencies and institutions. Interventions undertaken by some programmes and institutions, such as transaction structuring and co-investing, require different skills than providing subsidies and concessional lending. In addition, such interventions may cover a number of

unrelated sectors. As a result, bringing these functions together in the same institution may not yield efficiency gains. However, efficiency gains could result from bringing together transactional expertise in similar technologies, projects and business models, particularly if staff have the financial and sector knowledge to undertake a range of interventions. Consolidation of programmes and related outreach would also facilitate information sharing with retail and commercial customers and other investors (CPI, 2015).

Publication overview

This report is divided into five chapters. The remaining chapters will address the following topics:

- Chapter 2 examines GIB strategic investment mandates including the importance of profitability, project replicability and demonstrating that profitable investments are possible. It also provides an overview of the range of GIB target sectors and sub-sectors.
- Chapter 3 discusses the specific instruments and funds GIBs use to make investments. It draws attention to the range of de-risking approaches used by GIBs and the innovative approaches they are using to reduce high transaction costs. The chapter also discusses the types of co-investors that GIBs collaborate with or seek to attract.
- Chapter 4 explores how GIBs are mobilising private investment in domestic energy efficiency.
- Chapter 5 provides practical information on capitalising and setting up a GIB. Administrative set up, leadership and staffing are discussed, as well as reporting, oversight and transparency.

Notes

1. The Montgomery County Green Bank (Maryland, United States) is not included in the table. As of 10 May 2016, the Green Bank was just beginning the process of recruiting its board of directors.
2. A private equity fund is a fund which invests its money in an asset class consisting of equity securities in operating companies that are not publicly traded on the stock exchange, to control the company.
3. This topic has been the focus of extensive OECD analysis www.oecd.org/env/cc/financing.htm www.oecd.org/env/cc/financing.htm.
4. In financial management, "pure play" entities are focused on only one industry or product.

5. In March 2015, the UK DECC and the UK Green Investment Bank announced a pilot joint venture to deploy capital from the United Kingdom's International Climate Fund (ICF). The joint venture, named UK Climate Investments, focuses on renewable energy and energy efficiency in developing countries, including India, South Africa and countries in East Africa. The investment approach follows the UK Green Investment Bank business model and focuses on investing in green projects on commercial terms and mobilising private sector investment. A dedicated team manages the project and is supervised by a Board with members from the DECC and the UK Green Investment Bank (UK Green Investment Bank, 2015a; UK House of Commons, 2015, UK Green Investment Bank, 2015b).

References

Ang, G. and V. Marchal (2013), "Mobilising private investment in sustainable transport: The case of land-based passenger transport infrastructure", *OECD Environment Working Papers*, No. 56, OECD Publishing, Paris, http://doi.org/10.1787/5k46hjm8jpmv-en.

CCICED (2015), "Green financial reform and green transformation", report to the Annual Conference of CCICED, 9-11 November, China Council for International Cooperation on Environment and Development, www.cciced.net/encciced/policyresearch/report/201511/P020151117574533056430.pdf.

Corfee-Morlot, J. et al. (2012), "Towards a green investment policy framework: The case of low-carbon, climate-resilient infrastructure", *OECD Environment Working Papers*, No. 48, OECD Publishing, Paris, http://dx.doi.org/10.1787/5k8zth7s6s6d-en.

CPI (2015), "Getting the most from your green: An approach to using public money effectively through green banks and other low-carbon financing", Climate Policy Initiative, http://climatepolicyinitiative.org/wp-content/uploads/2015/12/Getting-the-Most-from-Your-Green-An-Approach-to-Using-Public-Money-Effectively-Through-Green-Banks-and-Other-Low-Carbon-Financing.pdf.

CPI (2014), "The global landscape of climate finance 2014", CPI Report, Climate Policy Initiative, November, http://climatepolicyinitiative.org/wp-content/uploads/2014/11/The-Global-Landscape-of-Climate-Finance-2014.pdf.

CPI (2013), "The global landscape of climate finance 2013", CPI Report, Climate Policy Initiative, October, www.climatepolicyinitiative.org/wp-content/uploads/2013/10/The-Global-Landscape-of-Climate-Finance-2013.pdf.

Eklin, K. et al. (2016, forthcoming), "OECD Green Investment Financing Forum: Lessons from established and emerging green investment bank models", Background Note, OECD, Paris, forthcoming.

ERB (2015), "New Jersey Energy Resilience Bank Grant and Loan Financing Program Guide", revised 15 October, New Jersey Energy Resilience Bank, www.njeda.com/pdfs/ERB/ERB_ProgramGuide_10_15_15.aspx.

Global Commission on the Economy and Climate (2014), *Better Growth, Better Climate: The New Climate Economy Report*, Global Commission on the Economy and Climate, http://newclimateeconomy.report.

Green Investment Bank Commission (2010), "Unlocking investment to deliver Britain's low carbon future", Green Investment Bank Commission, London, www.e3g.org/docs/Unlocking_investment_to_deliver_Britains_low_carbon_future_-_Green_Investment_Bank_Commission_Report_June_2010.pdf.

Haščič, I. et al. (2015), "Public interventions and private climate finance flows: Empirical evidence from renewable energy financing", *OECD Environment Working Papers*, No. 80, OECD Publishing, Paris, http://dx.doi.org/10.1787/5js6b1r9lfd4-en.

IEA (2014), *World Energy Investment Outlook*, OECD/IEA, Paris, www.iea.org/publications/freepublications/publication/WEIO2014.pdf.

IDFC (2015), "IDFC green finance mapping for 2014", International Development Finance Club, December, http://www.idfc.org/Downloads/Publications/01_green_finance_mappings/IDFC_Green_Finance_Mapping_Report_2015.pdf.

Kennedy, C. and J. Corfee-Morlot (2012), "Mobilising investment in low carbon, climate resilient infrastructure", *OECD Environment Working Papers*, No. 46, OECD Publishing, Paris, http://dx.doi.org/10.1787/5k8zm3gxxmnq-en.

KfW (2016), "KfW at a glance: Facts and figures", KfW Group, April, www.kfw.de/PDF/Download-Center/Konzernthemen/KfW-im-%C3%9Cberblick/KfW-an-overview.pdf

New Jersey Board of Public Utilities (2014), "Christie administration takes steps to establish Energy Resilience Bank", News Release, 23 July, www.state.nj.us/bpu/newsroom/announcements/pdf/20140723erb.pdf.

OECD (2015a), *Mapping Channels to Mobilise Institutional Investment in Sustainable Energy*, Green Finance and Investment, OECD Publishing, Paris, http://dx.doi.org/10.1787/9789264224582-en.

OECD (2015b), *Policy Guidance for Investment in Clean Energy Infrastructure: Expanding Access to Clean Energy for Green Growth and Development*, OECD Publishing, Paris, http://dx.doi.org/10.1787/9789264212664-en.

OECD (2015c), *Aligning Policies for a Low-carbon Economy*, OECD Publishing, Paris, http://dx.doi.org/10.1787/9789264233294-en.

OECD (2013a), "Long-term investors and green infrastructure", Policy Highlights brochure, OECD, Paris, www.oecd.org/env/cc/Investors%20in%20Green%20Infrastructure%20brochure%20(f)%20[lr].pdf.

OECD (2013b), *OECD Investment Policy Reviews: Malaysia 2013*, OECD Publishing, Paris, http://dx.doi.org/10.1787/9789264194588-en.

REN21 (2015), *Renewables 2015: Global Status Report*, www.ren21.net/wp-content/uploads/2015/07/GSR2015_KeyFindings_lowres.pdf.

Smallridge, D. et al. (2013), "The role of national development banks in catalyzing international climate finance", Inter-American Development Bank, Washington, DC, March, http://publications.iadb.org/handle/11319/3478?locale-attribute=en.

UK Green Investment Bank (2015a), "UK Green Investment Bank agrees to terms on £200m international pilot", press release, 24 March, www.greeninvestmentbank.com/news-and-insight/2015/uk-green-investment-bank-agrees-terms-on-200m-international-pilot.

UK Green Investment Bank (2015b), "UK Climate Investments launched with £200m to invest in India and Africa", press release, 13 November, http://www.greeninvestmentbank.com/news-and-insight/2015/uk-climate-investments-launched-with-200m-to-invest-in-india-and-africa/.

UK House of Commons (2015), "House of Commons: Written Statement (HCWS462)", Department for Energy and Climate Change, Written Statement made by Secretary for State for Energy and Climate Change (Mr. Edward Davey) on 24 Mar 2015, www.parliament.uk/documents/commons-vote-office/March%202015/24%20March/13-DECC-Investments.pdf.

UK House of Commons (2011), "The Green Investment Bank", Second Report of Session 2010-11, Volume I: Report, together with formal minutes, oral and written evidence, 11 March, House of Commons, The Stationary Office, London, www.publications.parliament.uk/pa/cm201011/cmselect/cmenvaud/505/505.pdf.

UNFCCC (2015), "Report of the Conference of the Parties on its twentieth session, held in Lima 1 to 14 December 2014: Addendum: Part two: Action taken by the Conference of the Parties at its twentieth session", FCCC/CP/2014/10/Add. 1, United Nations Framework Convention on Climate Change, http://unfccc.int/documentation/documents/advanced_search/items/6911.php?priref=600008358.

US EIA (2012), "State electricity profiles – Data for 2012 (released May 1, 2014)", United States Energy Information Administration, www.eia.gov/electricity/state.

Annex 1.A1.

Other institutions, funds, agencies and programmes that mobilise private climate finance and investment

Overview of entities engaged in mobilising private investment in low-carbon and climate-resilient infrastructure

The following types of public financial institutions, funds, government programmes and agencies are active in mobilising private investment in low-carbon and climate-resilient (LCR) infrastructure, and are distinct from green investment banks (GIBs) or GIB-like entities for the purpose of this report. This list is intended to highlight many of the major entities without attempting to be comprehensive. These institutions and entities include the following:

- Green programmes or initiatives within existing national development banks, bilateral development finance institutions, export credit agencies, multilateral development banks or central banks. These actors may invest domestically, internationally or both. National development banks and other types of public financial institutions (PFIs) are discussed at length in the OECD working paper "Public financial institutions and the low-carbon transition: Five case studies on low-carbon infrastructure and project investment" (Cochran et al., 2014). For example, the Brazilian development bank BNDES funds renewable energy, energy efficiency and public transport projects in Brazil through its Green Economy programme (BNDES, 2013) and the German development bank KfW has invested in environmental protection domestically and internationally since the 1980s (Cochran et al., 2014). Central banks can also lead or participate in green finance initiatives, as illustrated by Bangladesh Bank's establishment of a "Green Transformation Fund" to support green practices in the export-oriented textiles and leather sectors (Bangladesh Bank, 2016).

  PFIs are typically active in sectors where market failures have substantially limited private sector investment, and often hold a mandate to provide long-term financing independent of market cycles and in line with policy-oriented objectives. PFIs are able to leverage capital at advantageous, below-market rates for targeted investments. In many instances these institutions serve as a catalyst for private sector investment and innovation. These characteristics and objectives of PFIs are well aligned with the challenge of overcoming barriers to private investment in low-carbon projects. Some PFIs already have an explicit mandate and authority to invest in green infrastructure – often with established guidelines on which technologies or markets to address (Cochran et al., 2014).

  In the context of developing strategies to mobilise private investment in domestic green infrastructure, countries with an existing national development bank (or a public investment, infrastructure or industrial development bank which focuses on domestic investment) should consider and assess how its current operations are approaching green infrastructure investment, and the institution's relative success in scaling up and attracting private investment. "Greening" existing PFIs, when the necessary institutional and political support exists, might be preferable to creating new institutions.

- The multitude of **government-sponsored loan programmes** that provide financing for LCR projects are considered not to be GIBs or GIB-like entities. These programmes typically have less independence and flexibility than GIBs, although they may to different extents share with GIBs a focus on preservation and recycling of public capital. Government-sponsored loan programmes may provide significant amounts of grants, finance the majority of a project or provide non-commercial (i.e. subsidised) lending terms. One such programme is the Environmental Investment Fund of Namibia, which is funded by national conservation fees and environmental taxes and which provides subsidised loans and grants for sustainable resource management and environmental protection, including renewable energy, water efficiency and land-use planning (EIF Namibia, n.d.). South Africa's Green Fund, which focuses on green cities and towns, the low-carbon economy, and environmental and natural resource management, has similarities to a GIB-like entity but is focused on grants and has not mobilised significant private investment to date (Green Fund, 2014; personal communication with Ruan Kruger, Development Bank of South Africa, 25 November 2015).

- **Agencies and institutions supporting research and development (R&D) and early development of clean technology** (i.e. "clean tech" – a broad category which includes technologies that use less energy, generate less waste and cause less environmental damage). Publicly and privately supported research and industrial agencies encourage fundamental research in pre-commercial technologies, while GIBs generally operate at the other end of the spectrum, mobilising private investment for commercially established or market-ready technologies. Some organisations support business and product development for pre-commercial technologies. For example, Sustainable Development Technology Canada (SDTC), an independent non-profit funded by the government of Canada, supports the early-stage development of clean tech projects through funding and coaching in areas such as climate change, air quality and biofuels (SDTC, 2015).

- **Public agencies that implement national energy plans** are often designed to encourage private investment yet often lack the flexibility and independence of a GIB. Examples include the Moroccan Agency for Solar Energy (MASEN) and the Indian Renewable Energy Development Agency (IREDA).

- **Multilateral infrastructure development banks and other infrastructure-specific initiatives** are primarily concerned with scaling up and prioritising infrastructure investment without specifically focusing on LCR infrastructure. Box 1.A1.1 discusses some of these emerging infrastructure banks and initiatives.

- **Purely international funds** are also outside the scope of this report, as a key characteristic of GIBs is their focus on mobilising investment in *domestic* infrastructure. For example, the Danish Climate Investment Fund uses an innovative structure and risk mitigants to mobilise private investment (including Danish institutional investment) for projects in LCR infrastructure in developing countries.

The following sub-section provides additional details on selected international climate funds.

Box 1.A1.1. Emergence of new development banks and initiatives focused on infrastructure investment

Green investment banks (GIBs) are not the only institutions with significant new activity in financing infrastructure investment in recent years. Developing countries and emerging economies are undertaking efforts to establish new development banks. The Group of Twenty (G20) is also promoting increased infrastructure investment through a Global Infrastructure Initiative launched in 2014. While these institutions will not be exclusively dedicated to financing low-carbon and climate-resilient (LCR) infrastructure, they are particularly focused on filling the infrastructure gap.

In July 2014, Brazil, the Russian Federation, India, the People's Republic of China and South Africa (known as the BRICS countries) collectively agreed at their annual BRICS Summit to establish a development bank, to be known as the New Development Bank. The bank's purpose is to increase economic co-operation between BRICS countries and finance investment in sustainable development and infrastructure in BRICS and other emerging economies and developing countries (BRICS Summit, 2014). The New Development Bank "shall support public and private projects through loans, guarantees, equity participation and other financial instruments" and will also co-operate with other international organisations (BRICS Summit, 2014). The New Development Bank is based in Shanghai and is capitalised with the payment of USD 10 billion for each founding member for a total initial capitalisation of USD 50 billion (Hou, 2014). In April 2016, the bank approved its first set of loans for a total of USD 811 million supporting 2,370 MW of renewable energy capacity (New Development Bank, 2016). In addition, China has proposed an Asian Infrastructure Investment Bank (AIIB), which will provide project loans to developing countries. The agreement to establish the AIIB was signed by representatives from 50 countries in June 2015. The bank declared itself open for business on 16 January 2016 (AIIB, 2016).

Leaders of the G20 are increasing support to infrastructure investment through the creation of a Global Infrastructure Hub, which will provide resources to help implement the agenda of the Global Infrastructure Initiative. The Global Infrastructure Hub will have a four-year mandate to increase knowledge-sharing, address data gaps relevant for investors, increase the capacity of government officials and enhance investment opportunities by developing a database of infrastructure projects (G20, 2014). The Business 20 (B20), a forum of private sector leaders that produces policy recommendations for the G20, estimates that the Global Infrastructure Initiative can help to unlock an additional USD 2 trillion in global infrastructure capacity, 10 million jobs per annum and USD 600 billion in GDP benefits to 2030 (B20, 2014).

Sources: AIIB (2016), "AIIB open for business. Jin Liqun elected as first President", press release, 16 January, Asian Infrastructure Investment Bank, www.aiib.org/html/2016/NEWS_0116/84.html; AIIB (2015), "Fifty countries sign the articles of agreement for the Asian Infrastructure Investment Bank", press release, Asian Infrastructure Investment Bank, http://219.237.194.234/html/2015/NEWS_0629/11.html; B20 (2014), "B20 Infrastructure and Investment Task Force policy summary", B20 Infrastructure and Investment Task Force, July, www.b20australia.info/Documents/B20%20Infrastructure%20and%20Investment%20Taskforce%20Report.pdf; BRICS Summit (2014), "Agreement on the New Development Bank", VI BRICS Summit, Fortaleza, Brazil, 15 July, http://brics6.itamaraty.gov.br/media2/press-releases/219-agreement-on-the-new-development-bank-fortaleza-july-15; Hou, Z. (2014), "BRICS Development Bank, too good to be true?", Overseas Development Institute, 6 August, www.odi.org/comment/8703-brics-development-bank-too-good-be-true, G20 (2014), "The G20 Global Infrastructure Initiative", *G20 Communiqué*.

International climate funds and initiatives

A wide range of international climate funds and initiatives support greater financial mobilisation for mitigation and adaptation in developing countries.[1] Similarities between these funds and initiatives and GIBs vary depending upon their particular mandate and approach. The following non-exhaustive list highlights some of the more prominent international climate funds. Notably, this list does not include public finance from bilateral finance institutions and providers, which exceeded multilateral public finance in 2013 and 2014 (OECD, 2015).[2]

Green Climate Fund

The Green Climate Fund (GCF) is a fund created within the framework of the UNFCCC and is an operating entity of the financial mechanism of the Convention (UNFCCC, 2010). Operational since 2014, the GCF is designed to be used as a mechanism to disburse international climate finance provided mainly (but not exclusively) by developed countries to fund projects in developing countries. In November 2015, it approved its first investments for a total of USD 168 million in 8 projects and programmes valued at USD 624 million (GCF, 2015a). As of April 2016, the GCF was capitalised with USD 9.9 billion signed out of a total amount announced of USD 10.3 billion, with pledges from 42 countries, including 9 developing countries (GCF, 2016).

The GCF also seeks to increase complementarities between its activities and those of other relevant institutions. GIBs appear to be one type of institution whose activities are consistent with and supportive of the GCF's objectives. In terms of sectoral coverage, GIB investment activity is well aligned with the GCF; GIBs finance projects in three mitigation sectors identified as priorities by the GCF: renewable energy, energy efficiency and transport. Perhaps most importantly for the purposes of this report, the GCF includes a Private Sector Facility which will operate as a component of the fund.

To mobilise private capital and expertise at scale in accordance with national plans and priorities, the Private Sector Facility will address barriers to private sector investment in adaptation and mitigation activities, such as market failures, insufficient capacity and lack of awareness. These activities are expected to facilitate and enhance the participation of national, regional and international investors. GIBs, by their very nature, are designed to interact with private investors, which may be local or international, and to maximise their investment in green investment bank-supported projects and programmes. Lessons learnt by GIBs in engaging with institutional investors such as pension funds and investment funds may also be particularly useful for the Private Sector Facility. The GCF Private Sector Facility will seek to mobilise institutional capital from local actors and in the immediate term will focus on mobilising funds at scale from local commercial banks, local pension funds, local insurance companies, sovereign wealth funds and high net worth individuals (GCF, 2015b). Given that GIBs often partner with local banks (see the discussion on co-investors in Chapter 3), they are particularly well placed to share lessons learnt regarding collaboration with local banks and other investors.

Climate Investment Funds

The Climate Investment Funds (CIF) were developed in 2008 and designed as a way to mobilise resources to support climate change in developing countries in the areas of clean technology, renewable energy, sustainable forestry management and climate resilience. With a total capitalisation of USD 5.5 billion, the Clean Technology Fund (CTF) is one of the largest multilateral mitigation funds and one of the key funding areas of the CIF (UNFCCC, 2014a). The CTF provides concessional resources to scale up demonstration, deployment and technology transfer of low-carbon technologies (CIF,

2015). GIBs that support the development of clean technology, such as the Swiss Technology Fund and Masdar, have relevant expertise that could be shared.

Global Environment Facility

Originally established as a USD 1 billion World Bank pilot programme in 1991, the Global Environment Facility (GEF) has become a permanent and separate institution which serves as a financial mechanism for several international conventions, including the UNFCCC, the Convention on Biological Diversity and the UN Convention to Combat Desertification (GEF, 2013). In the context of the UNFCCC, the GEF supports projects in climate change mitigation and adaptation and administers the GEF Trust Fund, Least Developed Countries Fund and Special Climate Change Fund (SCCF). These funds provide grant support and lending for mitigation and adaptation projects. The SCCF, in particular, focuses on adaptation and technology transfer with a strong demand for projects related to water resource management resilience. GIB-like entities such as the New Jersey Energy Resilience Bank, Swiss Technology Fund or Masdar share some characteristics with the SCCF, and these institutions could potentially benefit from sharing relevant expertise and experience.

Adaptation Fund

The Adaptation Fund (AF), which was established to finance adaptation projects in developing countries that are particularly vulnerable to the impacts of climate change, has been operational since 2009. The AF is funded by a 2% levy on certified emission reductions issued to clean development mechanism projects as well as voluntary contributions (UNFCCC, 2014b). The AF is administered by the Adaptation Fund Board. As of December 2015 the AF has allocated USD 331 million and disbursed USD 140.6 million (UNFCCC, 2015). GIBs that support adaptation and resiliency may have relevant expertise to share with the AF, and vice versa.

Notes

1. The OECD's *Climate Fund Inventory Database* (http://qdd.oecd.org/subject.aspx?subject=climatefundinventory) is an available source that compiles funds for both adaptation and mitigation.

2. The OECD report "Climate finance in 2013-14 and the USD 100 billion goal" (OECD, 2015), prepared in collaboration with Climate Policy Initiative, provides an aggregate estimate of mobilised climate finance and an indication of the progress towards the UNFCCC climate finance goal, covering international climate funds and other sources of climate finance.

References

AIIB (2016), "AIIB open for business. Jin Liqun elected as first President", press release, 16 January, Asian Infrastructure Investment Bank, www.aiib.org/html/2016/NEWS_0116/84.html.

AIIB (2015), "Fifty countries sign the articles of agreement for the Asian Infrastructure Investment Bank", press release, Asian Infrastructure Investment Bank, http://219.237.194.234/html/2015/NEWS_0629/11.html.

B20 (2014), "B20 Infrastructure and Investment Task Force policy summary", B20 Infrastructure and Investment Task Force, July, www.b20australia.info/Documents/B20%20Infrastructure%20and%20Investment%20Taskforce%20Report.pdf.

Bangladesh Bank (2016), "Green finance for sustainable development", speech by Dr. Atiur Rahman, Governor, Bangladesh Bank to the Green Finance for Sustainable Development Conference, 12 January, www.bb.org.bd/governor/speech/jan122016gse732.pdf.

BNDES (2013), *Annual Report 2013*, BNDES – The Brazilian Development Bank, www.bndes.gov.br/SiteBNDES/export/sites/default/bndes_en/Galerias/RelAnualEnglish/ra2013/Rel_Anual_2013_ingles.pdf.

BRICS Summit (2014), "Agreement on the New Development Bank", VI BRICS Summit, Fortaleza, Brazil, 15 July, http://brics6.itamaraty.gov.br/media2/press-releases/219-agreement-on-the-new-development-bank-fortaleza-july-15.

CIF (2015), "About the Climate Investment Funds", Climate Investment Funds website, www.climateinvestmentfunds.org/cif/aboutus (accessed 1 April 2015).

Cochran, I. et al. (2014), "Public financial institutions and the low-carbon transition: Five case studies on low-carbon infrastructure and project investment", *OECD Environment Working Papers*, No. 72, OECD Publishing, Paris, http://dx.doi.org/10.1787/5jxt3rhpgn9t-en.

EIF Namibia (n.d.), Environmental Investment Fund of Namibia website, www.eifnamibia.com (accessed 21 August 2015).

G20 (2014), "The G20 Global Infrastructure Initiative", *G20 Communiqué*.

GCF (2016), "Status of pledges and contributions made to the Green Climate Fund", *Pledge Tracker*, Green Climate Fund, 20 April, www.greenclimate.fund/documents/20182/24868/Status+of+Pledges+%282016.4.20%29.pdf/eef538d3-2987-4659-8c7c-5566ed6afd19.

GCF (2015a), "Green Climate Fund approves first 8 investments", press release, 6 November, Green Climate Fund, www.greenclimate.fund/documents/20182/38417/Green_Climate_Fund_approves_first_8_investments.pdf/679227c6-c037-4b50-9636-fec1cd7e8588.

GCF (2015b), "Private Sector Facility: Potential approaches to mobilizing funding at scale", GCF/B.09/11, Meeting of the Board, 24-26 March 2015, Songdo, Korea, www.gcfund.org/fileadmin/00_customer/documents/MOB201503-9th/11_-_Potential_Approaches_to_Mobilizing_Funding_at_Scale_20150218_fin.pdf.

GEF (2013), "What is the GEF", Global Environment Facility website, www.thegef.org/gef/whatisgef (accessed 1 April 2015).

Green Fund (2014), "2013/2014 Green Fund annual report", Department Environmental Affairs, Republic of South Africa, www.sagreenfund.org.za/wordpress/wp-content/uploads/2014/12/GF-Annual-Report-2013-2014-Website.pdf.

Hou, Z. (2014), "BRICS Development Bank, too good to be true?", Overseas Development Institute, 6 August, www.odi.org/comment/8703-brics-development-bank-too-good-be-true.

New Development Bank (2016), "First set of loans approved by the board of directors of the New Development Bank", press release, 15 April, http://www.ndb.int/first-set-of-loans-approved-by-the-board-of-directors-of-the-new-development-bank.php#parentHorizontalTab2.

OECD (2015), "Climate finance in 2013-14 and the USD 100 billion goal", a report by the Organisation for Economic Co-operation and Development (OECD) in collaboration with Climate Policy Initiative (CPI), www.oecd.org/environment/cc/OECD-CPI-Climate-Finance-Report.pdf.

SDTC (2015), "Our approach", Sustainable Development Technology Canada, www.sdtc.ca/en/about-sdtc/our-approach.

UNFCCC (2015), "Report of the Adaptation Fund board", United Nations Framework Convention on Climate Change, http://unfccc.int/resource/docs/2015/cmp11/eng/02.pdf.

UNFCCC (2014a), "Green Climate Fund", United Nations Framework Convention on Climate Change, http://unfccc.int/cooperation_and_support/financial_mechanism/green_climate_fund/items/5869.php.

UNFCCC (2014b), "UNFCCC Standing Committee on Finance: 2014 biennial assessment and overview of climate finance flows", United Nations Framework Convention on Climate Change, Bonn, Germany, http://unfccc.int/files/cooperation_and_support/financial_mechanism/standing_committee/application/pdf/2014_biennial_assessment_and_overview_of_climate_finance_flows_report_web.pdf.

UNFCCC (2010), "Report of the Conference of the Parties on its sixteenth session, held in Cancun 29 November to 10 December 2010: Addendum: Part two: Action taken by the Conference of the Parties at its sixteenth session", FCCC/CP/2010/7/Add. 1, United Nations Framework Convention on Climate Change, http://unfccc.int/resource/docs/2010/cop16/eng/07a01.pdf.

Chapter 2.

Green investment bank mandates and target sectors

This chapter examines the investment mandates and target sectors and sub-sectors of green investment banks. It addresses the range of green investment banks' investment objectives including the importance of profitability, leveraging additional private investment and demonstrating that commercial investments are possible. The chapter also discusses other mandates such as promoting project replicability and encouraging standardisation. It closes with a discussion of the range of sub-sectors that green investment banks target for investment.

Key takeaways

- Governments have created green investment banks with a variety of strategic mandates to address market barriers and meet national and sub-national objectives.
- Most green investment banks focus on promoting investment in renewable energy and energy efficiency but some entities target broader areas, such as innovation, resilience or sustainable cities.
- Green investment banks are often mandated to provide "additional" capital and must demonstrate "additionality" – i.e. that their participation in a transaction attracted private capital that otherwise would not have been invested.
- In some jurisdictions, green investment banks are required to be, and are, profitable. Achieving this objective can increase political support for dedicating public resources to mobilise private investment in climate change mitigation and resilience.
- Many green investment banks aim to attract private capital to underserved, yet commercially viable, low-carbon and climate-resilient infrastructure markets by demonstrating to the private sector how deals can be structured and risks reduced.

Strategic investment mandates of green investment banks

Green investment banks (GIBs) are mandated by governments to fulfil specific objectives. These objectives may be straightforward, such as deploying a given amount of capital into a specific sector such as energy efficiency, but they are often broader and allow for flexibility to address particular challenges and objectives. For example, NY Green Bank seeks to invest where there is a financing gap and focuses on "clean energy projects that are economically viable but not currently financeable" (NY Green Bank, 2013). This chapter discusses different GIB mandates, including profitability and leveraging of private capital, demonstration, additionality, replicability and standardisation. It also explores the range of GIB target sectors and sub-sectors. Chapter 3 discusses the actual investment instruments, funds and de-risking approaches GIBs use to execute their mandates.

Profitability, and leveraging and recycling public capital

Unlike many grant-making public institutions, GIBs seek to recycle public capital and focus on mobilising private investment using public capital. Some GIBs are also required to be, and are, profitable. For example, the UK Green Investment Bank must invest on commercial terms, and has to meet a minimum 3.5% annual nominal return on total investments, after operating costs but before tax (UK Green Investment Bank, 2015a). It describes itself as "unashamedly and unambiguously a for-profit bank" (UK Green Investment Bank, 2013a). This objective is part of the UK Green Investment Bank's "double bottom line" in which its green impact and financial results are considered equally important performance indicators.

Australia's Clean Energy Finance Corporation (CEFC) is required to compare its financial performance with a portfolio benchmark return (CEFC, 2014a). In February 2015, the Australian government altered the CEFC's mandate to increase its average return to at least the five-year Australian government bond rate plus 4-5% per annum, before operating costs (Australian Treasury, 2015). The CEFC stated that this target is "highly challenging" and that achieving such increased returns without increasing risk is likely to be "outside the scope of normal market opportunities" (CEFC, 2015a). In December 2015, the CEFC received a new Investment Mandate, which modified the investment portfolio risk profile to accommodate an increased focus on emerging and innovative renewable technologies, energy efficiency and energy efficiency technologies for cities and the built environment (Commonwealth of Australia, 2015).

The decision to set profitability targets for GIBs has implications for and is reflected in their mandates and strategies. The UK Green Investment Bank states that all of its investments have "been made on fully commercial terms with returns in line with other private sector investors (UK Green Investment Bank, 2015a)." Australia's CEFC is committed to investing on commercial terms and achieving financial self-sustainability to counter market failures and financing impediments and to generate positive public policy outcomes.[1] Nevertheless, it can also offer concessional financing, subject to an AUD 300 million per year limit (CEFC, 2013). Similarly, investments by the Connecticut Green Bank are generally market based. However, in cases where it is required by statute to promote or support a particular technology, such as combined heat and power (CHP) or anaerobic digestion, the Connecticut Green Bank has provided concessional financing (which ultimately can be recycled) as an alternative to outright grants (CEFIA, 2014; personal communication with Bert Hunter, Connecticut Green Bank, 7 January 2016).

GIBs seek to multiply their impact by leveraging significant private investment for every unit of public capital spent, and they measure this impact through "leverage ratios" (i.e. private investment mobilised per unit of GIB public spending). For example, the UK Green Investment Bank has mobilised an estimated GBP 3 of private capital for every GBP 1 of public investment it has made since its inception (UK Green Investment Bank, 2015a). The Connecticut Green Bank attracted USD 10 in private investment for every USD 1 of public capital spent in 2013 (Connecticut Green Bank, 2013). In 2014, the ratio was USD 3 of private investment for every USD 1 of private capital spent (Connecticut Green Bank, 2015a). The CEFC reported AUD 1.8 private dollars mobilised for each AUD 1 in CEFC investment in 2014-15 (CEFC, 2015b; 2015c). In 2013-14, this ratio was AUD 2.2:1 (CEFC, 2014a). NY Green Bank estimates that its first group of transactions will leverage up to USD 3.6 private dollars per public dollar (NY Green Bank, 2015a).

In light of potential differences among leverage calculation methodologies, caution should be used when comparing mobilisation rates across institutions. Caruso and Ellis (2013) examined how 24 actors involved in climate financing defined and quantified the level of private climate finance mobilised by their interventions. They found a wide variation in the stringency of methods used to assess whether, and to what extent, climate finance has been mobilised with varying methodologies between and within different financial institutions and across financial instruments.

GIBs can also facilitate "capital recycling" by helping to refinance existing investments, thereby freeing up capital for future investments (UK Green Investment Bank, 2015a). For example, rather than "pushing" more financing into the construction phase, the UK Green Investment Bank works to "pull" more capital through the pipeline by refinancing operational-stage projects. By providing refinancing at that stage, GIBs provide project developers and other early-stage investors an "exit strategy", allowing them to free up capital to invest in new projects – i.e. to "recycle" their capital. During the operational stage, the majority of project-based risks have been resolved and projects typically produce consistent returns. At this point in the project cycle, different classes of institutional investors may be better able to meet their risk-return and liquidity expectations. The UK Green Investment Bank's Offshore Wind Fund, which has raised over GBP 600 million of private capital, demonstrates capital recycling in action (see Chapter 3). The Connecticut Green Bank has used a similar approach by establishing a USD 40 million facility for construction-phase lending and loan aggregation for commercial and industrial energy efficiency and renewable energy investments which are sold off to institutional investors, supporting the operational phase (Connecticut Green Bank, 2014).

Not all GIBs and GIB-like entities put an emphasis on financial returns. For example, Malaysia's Green Technology Financing Scheme (GTFS), which is administered by GreenTech Malaysia, does not have a mechanism to generate returns as it provides only loan guarantees and subsidised loans. This targeted approach may be compatible with GreenTech Malaysia's objective – developing sustainable and widespread green technology markets and strengthening the local green technology industry – but it restricts the institution's authority to pursue other approaches to mobilise investment. The New Jersey Energy Resilience Bank (ERB) offers financing paired with grants, with the opportunity for partial loan forgiveness (ERB, 2015). Given the urgency in increasing the resiliency of critical infrastructure, the ERB focuses on quickly scaling up resiliency investment and is not required to be profitable.

Demonstration

Many GIBs aim to attract private investment in low-carbon and climate-resilient (LCR) infrastructure sectors which have not yet received significant private investment. To do this, GIBs may choose to make investments more financially attractive by mitigating risk through credit enhancements. However, some GIBs believe that to achieve long-term, sustained private investment in a market sector, it is important to prove through "demonstration" that LCR infrastructure investments can be profitable today even without credit enhancements.

GIBs point out that some private investors simply may be unaware of LCR infrastructure investment opportunities, while others may recognise investment opportunities and understand risks but hesitate to enter the market due to lack of familiarity with deal structures and underwriting practices. GIBs can help to attract attention to these markets and can build an investment case over time using data on project performance and loan default rates. The Japanese Green Fund, for example, is particularly focused on providing accurate market information to private actors and showing that LCR investment models work.

Demonstration can also help address incorrect perceptions among investors that clean technologies are less developed, risky and not commercially viable. GIBs can demonstrate to (or with) private investors the actual risk profile of LCR infrastructure or highlight the significant market size and profit potential of a certain sector. GIBs also support both objectives through direct investment. By putting "skin in the game" (i.e. making actual investments, thereby demonstrating confidence in the underlying projects and a willingness to take on investment risks), the public sector signals that a project has a level of viability and certainty that can provide confidence to private investors to co-invest. For example, the UK Green Investment Bank focuses on demonstrating to investors that LCR investment opportunities can deliver commercially attractive returns even as it provides market-rate rather than subsidised financing.

Additionality

For many GIBs, "additionality" (or "financial additionality") is an important part of their mandate; the investment they attract must be *additional* to what would have otherwise occurred. Another expression closely linked to the concept of additionality is "crowding in" investment. Crowding in occurs when public investment induces greater private investment than would have occurred otherwise. "Crowding out" occurs when public intervention directly displaces private investment by undertaking projects the private sector would have otherwise financed. Crowding out can also occur indirectly if governments use taxes to fund public investment or in situations where demand for government borrowing results in increased interest rates, making borrowing too costly for private investors.

Additionality considerations for the UK Green Investment Bank are more particular. Because it is currently the recipient of government funding it is subject to European Union state aid rules. As a result, the UK Green Investment Bank must crowd in additional finance, rather than displacing other investors. More specifically, the UK Green Investment Bank must provide evidence that the projects it funds "have been denied funds or have not obtained all the necessary funding from market operators" (European Commission, 2012). In addition, the UK Green Investment Bank's interventions must abide by the "additionality principle", which dictates that UK Green Investment Bank funding, whenever possible, must come in addition to market financing (European Commission, 2012). In line with this requirement and its emphasis on demonstration and investing on commercial terms, the UK Green Investment Bank focuses on areas where its added "capital, knowledge and reputation can make the difference that enables a project to be financed successfully", rather than providing de-risking tools or concessional financing (UK Green Investment Bank, 2013b). While this may not solve all market "gaps", the UK Green Investment Bank believes that its approach will help keep private capital in the market after public financing or other support is removed (Cochran et al., 2014).

Additionality may be defined in different ways. In the view of NY Green Bank, the strictest interpretation of additionality would put the institution "one standard deviation away from the market frontier" (NY Green Bank, 2015b), where there is the most potential to attract and leverage private sector capital. NY Green Bank therefore takes a broader view on additionality and considers deals where it can bring a unique benefit to the proposed financing. For example, it can undertake investments that may occur in private markets but would involve less favourable terms, would lack the breadth needed to scale the market or would occur less quickly without NY Green Bank's involvement (NY Green Bank, 2015b). As noted in the discussion on whether GIBs are temporary or permanent institutions (see Chapter 5), several studies suggest that a lack of financial additionality is common among private sector investments mobilised by different types of public financial institutions (e.g. international development finance institutions, multilateral development banks). Given such concerns, additionality is a central focus of GIBs, as their ability to demonstrate that they are avoiding crowding out investment is an important factor in maintaining public support. Future research could focus on whether GIBs have been able to achieve their objective of crowding in (and not crowding out) private investment.

Replicability and standardisation

For practical and strategic purposes some GIBs strongly focus on project replicability. As NY Green Bank is mandated to support projects that will help transform the renewable energy financing market, it must demonstrate that any financing arrangements it enters into can be easily replicated and help achieve widespread deployment on a short-term basis (NY Green Bank, 2014a).

One other aspect of a GIB's business model that is frequently discussed in the US context is standardisation. The Coalition for Green Capital (a US-based non-governmental organisation which works to establish green banks) and NY Green Bank highlight the need for more standardised loan underwriting processes, contracts and data collection on loan and project performance. In their view, standardisation would make it easier and cheaper for securitisation to occur, for private banks to underwrite and for credit agencies to rate a securitisation.[2] The National Renewable Energy Laboratory, a part of the US government, has hosted the Solar Access to Public Capital working group of industry participants for several years to develop standardised power purchase agreements and related documentation for distributed solar projects, specifically to increase the flow of private capital to this market (NREL, 2013). In its analysis of channels for institutional investment in renewable energy infrastructure, the OECD has made recommendations to promote market transparency and standardisation, improve data, reduce transaction costs and support channels for securitisation of sustainable energy debt, including through supporting efforts to standardise contracts and project evaluation structures (OECD, 2015a).

Infrastructure resiliency

A GIB may choose to specifically focus on grid resiliency and infrastructure resiliency. The state of New Jersey (United States) created the Energy Resilience Bank (ERB) in response to Hurricane Sandy, which caused long-lasting power outages throughout the state (ERB, 2015). The ERB has limited its target market to key infrastructure such as hospitals and water facilities, and made only resilient technologies eligible for investment (e.g. energy storage and fuel cells). The ERB has also taken a different investment approach than other GIBs, offering a mix of low-cost capital and grants, while still seeking to leverage private investment. All of the ERB's projects require private investment, but grid resiliency represents an immediate, short-term requirement and the ERB is willing to sacrifice return on its capital in order to more rapidly draw investment into this critical area.

Innovation and technology

Most GIBs do not focus their support on research or early-stage technological development, but rather seek to invest primarily in proven commercial technologies. For example, NY Green Bank primarily supports technology deployment rather than technology development (New York Public Service Commission, 2013). This focus is generally more compatible with conserving public capital, but increases the risk that GIB investment is not additional. Despite this focus on deployment of existing technologies, GIBs typically retain flexibility to invest in new technologies that they think will be successful commercially after initial deals are facilitated by GIB interventions. GIBs' mandate to avoid crowding out private investment requires them to shift into new technologies with less attractive risk-return profiles when their interventions are no longer needed to attract investment, e.g. in sub-sectors and technologies in which risk-return profiles have improved due to cost declines and investment experience. The UK Green Investment Bank takes this approach and is monitoring carbon capture and storage, marine energy and transport biofuel technologies, but does not expect to make any near-term investments.

NY Green Bank notes that it is "technology risk averse" and will only deploy "commercially proven technologies" (NY Green Bank, 2014b). Yet being responsive to markets is a key part of NY Green Bank's mission to "transform financing markets" and its list of potential target technologies for investment is broad and includes ocean and tidal power, fuel cells and electric vehicle infrastructure (NY Green Bank, 2014a). Australia's CEFC believes that other government agencies are better placed to support early development and explicitly avoids investment in "early stage speculative technologies" as it considers such investment incompatible with its risk profile (CEFC, n.d.). The CEFC works with the Australian Renewable Energy Agency (ARENA), which supports technologies from early-stage research through to commercialisation by supporting the research, development and demonstration stages (Gray, 2013). However, the CEFC also retains flexibility to participate in transactions that require extra time and resources and that may not normally be justified in a purely profit-seeking institution. In addition, the CEFC can occasionally support projects such as a small and complex community wind farm deal, because the project creates a "public policy benefit" (CEFC, 2013).

More recently, the Australian government directed the CEFC to develop a focus on "emerging and innovative renewable technologies" in the Australian context such as "large-scale solar, battery storage associated with large and small scale solar and offshore wind" (Commonwealth of Australia, 2015). The lower threshold of acceptable financial return assigned to these projects can be balanced in the portfolio by other higher profit investments. To further promote innovation, in March 2016 the Australian government announced the creation of a new AUD 1 billion Clean Energy Innovation Fund. The fund will be jointly managed by the CEFC and the Australian Renewable Energy Agency and will provide both debt and equity for clean energy projects. It will focus on early-stage companies, business and projects seeking growth capital or early-stage capital (CEFC, 2016).

Like the CEFC, NY Green Bank takes a portfolio approach to investment and expects to support some innovation while also seeking immediate impact from mature renewable energy technologies (New York Public Service Commission, 2013). Similarly, the Connecticut Green Bank underscores that the mission of GIBs is not to support early-stage technologies, venture capital investments or research and development (R&D). However, when required by statute to promote or support a particular technology, such as CHP or anaerobic digestion, the Connecticut Green Bank has provided concessional financing as an alternative to grants.

Unlike most GIBs, the Swiss Technology Fund focuses primarily on innovation. Instead of identifying target technologies, the fund supports creditworthy and innovative companies that use novel techniques or products to reduce emissions or improve energy efficiency as well as technologies that have been tested and are market ready (Technology Fund, n.d.). Masdar also focuses on innovation. For example, the Masdar Clean Tech Fund is a USD 250 million venture capital fund focused on the development and commercialisation of a wide range of LCR technologies through investments in both projects and companies. In addition, Masdar Clean Energy, which is the GIB-like subsidiary of Masdar, invests in carbon capture and storage as part of its strategy to reduce emissions and promote technological development. Masdar Clean Energy is currently developing the Al Reyadah project, which will capture 800 000 tonnes of carbon dioxide (CO_2) annually. The project is a joint venture between Abu Dhabi National Oil Company (Adnoc) and Abu Dhabi Future Energy Company-Masdar and is expected to be operational in 2016 (Kader, 2015). Due to this support for early-stage technologies, Masdar Clean Energy can be considered to be a hybrid between a GIB and a technological support programme.

Target sectors and sub-sectors

GIBs are often required to invest in specific sectors or sub-sectors. Feasibility studies undertaken to inform the creation and design of GIBs typically consider the types of target sectors and interventions to pursue. For example, the UK government engaged Vivid Economics and McKinsey & Co. (2011) to study the sectors with the most significant market failures and capital shortages and the relative value for money resulting from investment in different sectors. The feasibility study highlighted offshore wind, energy efficiency and waste as key areas for action; these sectors eventually became the mandated target sectors for the UK Green Investment Bank. Other GIBs may be mandated broadly to invest in renewable energy with less focus on specific technologies. For example, the missions of NY Green Bank and the Connecticut Green Bank are specifically linked to promoting renewable energy investment.[3]

To date, GIBs such as Australia's CEFC, the Connecticut Green Bank, NY Green Bank and the UK Green Investment Bank, which are focused on profitability or financial sustainability, have mainly targeted renewable energy and energy efficiency. However, they also cover such technologies as alternative fuel vehicles and infrastructure (Connecticut Green Bank, CEFC), electric vehicle infrastructure (NY Green Bank) and waste recycling (UK Green Investment Bank). Other GIBs and GIB-like entities cover additional sectors and activities. Japan's Green Fund covers waste management; GreenTech Malaysia is allowed to support activities across the energy, water and waste treatment, building and transport sectors; and the Swiss Technology Fund can support natural resource conservation technologies. Table 2.1 shows target sectors and sub-sectors for operational GIBs and GIB-like entities. Specific target markets for GIB activity are described in the following section.

Renewable energy generation

Solar photovoltaic (PV) and onshore wind are the dominant renewable energy technologies in terms of new investment flows (OECD, 2015b). Solar PV and wind energy are also expected to account for the largest shares of estimated investment needs to achieve the 2°C target under the International Energy Agency's 2°C (2DS) scenario (29% and 23% respectively; IEA, 2012). Box 2.1 discusses the increased cost-competitiveness of renewable power and the potential role of GIBs to lower energy generation prices.

Renewable energy generation is a core focus for most GIBs. Although many GIBs have a mandate to support renewable energy broadly, most of their activities to date have been in the onshore wind and solar PV industries. Australia's CEFC, for example, must invest a minimum of 50% of its portfolio in renewable energy technologies by 2018; it met its target in 2014 and 2015 (CEFC, 2015c). Its portfolio in 2014-15 was dominated by solar PV (around 33%) and wind (21%), with other renewables such as solar thermal and bioenergy playing a smaller role (CEFC, 2015c). Solar PV programmes have been a core focus for the Connecticut Green Bank. The UK Green Investment Bank is unique in allocating a specific portion of its portfolio (and collectively, the majority of its portfolio) to designated renewable technologies (e.g. offshore wind and waste-to-energy).

Table 2.1. **Target sectors for green investment banks (GIBs) and GIB-like entities**

Entity	Target sectors and sub-sectors
California CLEEN Center (California, United States)	– Energy and water reduction and conservation projects for municipalities, universities, schools and hospitals – Clean electricity generation, distribution, transmission and storage – Energy conservation, environmental mitigation, and water treatment and distribution
Clean Energy Finance Corporation (Australia)	– Renewable energy (wind, solar PV, thermal and concentrated solar power, biomass, geothermal, tidal and other renewable energy [50%]) – "Low emissions" and energy efficiency (50%) – Within these target sectors, a focus on "emerging and innovative renewable technologies", energy efficiency and "energy efficiency technologies for cities and the built environment"
Connecticut Green Bank (Connecticut, United States)	– Energy efficiency – Renewable energy – Other clean technologies, including combined heat and power (CHP), anaerobic digestion, fuel cells, alternative fuel vehicles and infrastructure, storage and others
Green Energy Market Securitization (GEMS) (Hawaii Green Infrastructure Authority) (Hawaii, United States)	– Low- and moderate-income homeowners, renters and non-profits – Distributed solar PV (initial phase) – Clean energy and energy efficiency (deployed in phases)
Green Fund (Japan)	– Low-carbon projects (e.g. wind, solar, small-scale hydro, biomass, waste management, geothermal, hot springs, renewals of mid-sized hydro)
Malaysian Green Technology Corporation (GreenTech Malaysia) (Malaysia)	– Energy (renewable energy) – Water and waste management – Building (energy and water efficiency, indoor air quality)
Masdar (United Arab Emirates)	– Clean energy – Energy efficiency – Carbon capture and storage
New Jersey Energy Resilience Bank (ERB) (New Jersey, United States)	– CHP, fuel cells and solar-tied storage at water and wastewater treatment facilities
NY Green Bank (New York, United States)	– Energy efficiency – Renewable energy – Other clean technologies, including CHP, electric vehicle infrastructure, fuel cells and offshore wind
Technology Fund (Switzerland)	– Greenhouse gas reduction technologies – Energy efficiency – Renewable energy – Natural resource conservation technologies
UK Green Investment Bank (United Kingdom)	Priority areas: – Offshore wind – Waste recycling and bioenergy – Energy efficiency – Small-scale renewables Other: – Biofuels for transport, biomass power, carbon capture and storage, marine energy, renewable heat

Notes: Australia's CEFC will consider on a case-by-case basis the funding of "low-emissions technologies" which may include energy production and electricity generation using non-renewable fuels in cases where the investment will substantially lower current CO_2 emissions levels (CEFC, 2014b).

GIBs also support other renewable energy generation technologies like biomass and CHP. Some forms of thermal energy production may also attract GIB support. Solar thermal or heat pumps may provide more affordable and cleaner heating alternatives in regions that rely on inefficient carbon-based heating technologies. Box 2.2 highlights three GIB renewable energy investments.

Box 2.1. Falling price of renewable energy generation power and the potential role of green investment banks

The price of renewable energy generation has fallen dramatically in recent years. This decline is attributable to the fall in both the hard and soft costs of renewables. The hard cost decline refers to the falling price and increased efficiency of renewable technology itself. Wind turbines are increasingly efficient with relatively lower prices and the per-watt price of solar has fallen dramatically. Soft costs of renewables have also fallen. These include the cost of developing projects, marketing to customers and other required elements of completing a project unrelated to the technology.

The International Energy Agency (IEA) estimates that from 2010 to 2015, average costs for new onshore wind plants fell by 30% and average costs for new utility-scale solar photovoltaic (PV) installations declined by two-thirds (IEA, 2015). As of December 2015, contracted prices for PV-generated electricity were as low as USD 58/MWh[1] in the United Arab Emirates and USD 38.70/MWh[2] (escalating 3% per year) in Nevada (United States).[3] While prices vary significantly across regions and delivered project costs may differ from contracted costs, the IEA notes that the United Arab Emirates deal and bid and auction prices for solar PV and offshore wind in Brazil and South Africa "signal a step change in generation costs where deployment is starting to ramp up quickly" (IEA, 2015).

Some GIBs, such as the Connecticut Green Bank, seek to play a role in further reducing the price of renewable energy to make it competitive. The levelised cost of electricity from renewable energy is heavily dependent on the financing cost, as nearly all project costs are financed upfront. According to a study by Lazard, reducing the borrowing rate by 3.8 percentage points lowers the price of solar-powered electricity by 26%. Green investment banks can help bring down borrowing costs directly by offering capital at better terms, or indirectly through interventions which increase private lenders' confidence and willingness to offer lower cost debt. More private sector lending and increased understanding of the risks of renewables can reduce borrowing rates. Finally, by sharing data and standardising processes, GIBs can further reduce transaction costs associated with renewable power transactions.

Notes: 1. For a 25-year power purchase agreement (PPA) for electricity to be delivered by a 200 MW solar PV plant in 2017. 2. For a 20-year PPA for electricity to be delivered by a 100 MW solar PV plant by December 2016. 3. The levelised cost of electricity for the Nevada project is USD 48.61/MWh, which includes the business energy investment tax credit and network upgrade costs (Public Utilities Commission of Nevada, 2015; personal communication with Heymi Behar, IEA, 9 November 2015).

Sources: Shahan, Z. (2014), "13 charts on solar panel cost & growth trends", *Cleantechnica*, 4 September, http://cleantechnica.com/2014/09/04/solar-panel-cost-trends-10-charts; Lazard (2014), "Lazard's levelized cost of energy analysis -- Version 8.0, September", http://docplayer.net/30209-Lazard-s-levelized-cost-of-energy-analysis-version-8-0-september-2014.html; Meza, E. (2015), "US: World's highest prices for Chinese modules", *PV Magazine*, 2 March, www.pv-magazine.com/news/details/beitrag/us--worlds-highest-prices-for-chinese-modules-_100018400/#axzz3Y1g4m2f6; IEA (2015), *Medium-term Renewable Energy Market Report 2015*, OECD Publishing, Paris, http://dx.doi.org/10.1787/renewmar-2015-en; Kenning, T. (2015), "Buffett project's record low cost part of pricing 'trend', says First Solar", 9 July, PVTech, www.pv-tech.org/news/buffett_projects_record_low_cost_is_part_of_pricing_trend_says_first_solar.

Renewable energy storage

Due to the varying nature of renewable electricity generation, energy storage solutions can improve the attractiveness of renewable energy technologies. If electricity generated by sources such as solar and wind can be stored in batteries or other technologies, the total cost of renewable energy generation can be reduced and challenges of grid integration can be eliminated. For countries that lack robust transmission systems, renewable energy plus storage capability can remove the need to fund extensive transmission networks.

Box 2.2. Examples of green investment bank investments in renewable energy

Residential solar: NY Green Bank provided a USD 25 million warehouse credit facility to a New York-based solar provider that designs and installs systems for residential homes at no cost to the consumer. The project will demonstrate the commercial viability of underfunded, less well-known solar developers, which have more difficulty accessing financing than larger, better known developers.

Biogas: Australia's Clean Energy Finance Corporation (CEFC) provided an AUD 15 million market-rate loan to an Australian beef processor in order to build a biodigester on top of the processing facility. The loan helped the borrower receive commercial financing from its own private bank for the AUD 40 million project. The biodigester replaced a coal-fired power plant, and covers half of the facility's energy needs.

Waste-to-energy: A consortium comprising the UK Green Investment Bank, Balfour Beatty plc, Eternity Capital Management Limited, Foresight's UK Waste Resources and Energy Investments (UKWREI) Fund, in which the Green Investment Bank is a cornerstone investor, and the GCP Infrastructure Fund with the developer, Carbonarius invested GBP 47.8 million in a plant that will convert recovered wood into electricity using gasification technology. Over its expected 20-year lifetime the plant is forecast to supply enough renewable energy to power 17 000 homes each year and is expected to deliver emissions reductions of around 2.1 million tonnes of CO_2 equivalent, and to save around 1.3 million tonnes of wood from landfill. The UK Green Investment Bank directly invested GBP 12 million through preferred loan stock[1] and a further GBP 6.2 million in indirect investment through its cornerstone stake in UKWREI.

Notes: Preferred loan stock refers to stock shares used as collateral to secure a loan from another party. Preferred stocks have priority over common shares.

Sources: NY Green Bank (2015a), "Governor Cuomo announces three NY Green Bank transactions to improve access to clean energy and reduce greenhouse gas emissions", www.governor.ny.gov/news/governor-cuomo-announces-three-ny-green-bank-transactions-improve-access-clean-energy-and; UK Green Investment Bank (2013b), "Annual report 2013", Green Investment Bank, Edinburgh, www.gov.uk/government/uploads/system/uploads/attachment_data/file/336552/green-investment-bank-annual-report-2013.pdf; CEFC (2014c), "CEFC finance for Bindaree Beef biogas and rendering upgrade", Fact Sheet, Clean Energy Finance Corporation, Sydney, Australia, July, www.cleanenergyfinancecorp.com.au/media/76497/cefc-pdf-factsheet-bindaree_lr.pdf.

Some GIBs are investing in projects to promote renewable energy storage. For example, one of the first types of eligible projects for the New Jersey Energy Resilience Bank is energy storage tied to solar PV at water facilities (ERB, 2014). Private market participants, such as SolarCity, have already begun to offer residential rooftop solar PV paired with on-site batteries in pilot commercial programmes in the United States. GIBs can help expand these offerings, which would be particularly appealing in markets with high-priced grid electricity or poor track records for reliability.

Energy efficiency

Energy efficiency will play an important role in the transition to an LCR economy. Some of the barriers to both capital supply and consumer demand for energy efficiency investment can be addressed by GIBs. GIB activity and approaches for financing energy efficiency projects are discussed at length in Chapter 4. Box 2.3 highlights examples of energy efficiency investments by GIBs.

Box 2.3. Examples of green investment bank investments in energy efficiency

Commercial on-bill financing: Australia's Clean Energy Finance Corporation (CEFC) partnered with an Australian energy retailer to provide energy efficiency financing to commercial buildings to perform energy upgrades. The financing is paid back as an additional line item on a building's utility bill. This eases the repayment process for the borrower and also provides additional security to the lender, enabling longer terms and lower borrowing rates.

Municipal street lighting: The UK Green Investment Bank offers an innovative corporate loan facility to municipalities specifically tailored to allow cities to upgrade their street lighting to light emitting diodes (LEDs). This more efficient lighting technology saves more money for municipalities than the cost of the loan payment, allowing borrowers to be cash-flow positive throughout the period of the loan. With fixed rates and terms designed to match the payback period, municipalities are able to save 80% of their lighting costs by switching to LEDs.

Commercial PACE financing: The Connecticut Green Bank operates the state's Commercial Property-Assessed Clean Energy (C-PACE) loan programme. Through this structure, commercial building owners can receive long-term financing (up to 20 years) to perform energy upgrades on buildings, and pay the loan back as a new tax lien on the property. Linking the lien to the property increases the lending security, enabling a much longer payback term. The lien structure also makes it easier to buy and sell property with an outstanding energy efficiency loan. The Connecticut Green Bank originates loans and builds a portfolio, which is then sold to private investors to draw in private capital otherwise unable or unwilling to invest in smaller projects.

Sources: CEFC (2014d), "CEFC finances deep retrofits to unlock value in commercial property", press release, 28 March, www.cleanenergyfinancecorp.com.au/media/releases-and-announcements/files/cefc-finances-deep-retrofits-to-unlock-value-in-commercial-property.aspx; UK Green Investment Bank (2014), "Low energy streetlighting: Making the switch", UK Green Investment Bank, Edinburgh, February, www.greeninvestmentbank.com/media/5243/gib-market-report-low-energy-streetlighting-feb-2014-final.pdf; Connecticut Green Bank (2015b), "C-PACE marks successful first two years as CT property owners take advantage of program to finance money-saving energy improvements", press release.

Green cities

Some GIBs seek to support more sustainable and green urban environments. This can be done through a range of interventions, such as retrofitting buildings to be more energy efficient or building a more sustainable and ecologically friendly urban footprint. The UK Green Investment Bank's investment to support more efficient street lighting by city governments is another example of urban-focused LCR infrastructure investment (UK Green Investment Bank, 2015b). One of Masdar's core projects is Masdar City, a special economic zone outside of Abu Dhabi, which is a cluster for clean technology companies and powered entirely by renewable energy (Masdar, 2013). In 2015, the CEFC was tasked with an increased focus on "energy efficiency in the built environment" (Commonwealth of Australia, 2015) and launched an AUD 250 million fund targeting energy efficiency in local government (CEFC, 2015d).

Transportation

The transportation sector produces 23% of global fuel combustion-based CO_2 emissions in the world and is second only to the energy generation sector (which accounts for 42% of global fuel combustion-based CO_2 emissions) (IEA, 2014). Several GIBs include transport-related investments in their target sectors. For example, NY Green Bank has authority to invest in electric vehicle infrastructure and the UK Green Investment

Bank is approved to make investments in biofuels for transport. Australia's CEFC provides finance for vehicles assessed by Australia's regulators as being among the best performers in fuel efficiency and emissions reduction, with a concessional interest rate discount passed through to end users to encourage uptake of these vehicles (CEFC, 2015c; 2015d). A study by Frades, Peace and Dougherty (2014) proposes that GIBs could support the adoption of plug-in electric vehicles (PEVs) through direct finance for PEVs or electric vehicle supply equipment, creating partnerships with automobile equipment manufacturers, utilities and electric vehicle supply equipment installers. Any GIB activity in this area would need to be reconciled with the terms of its mandate (e.g. importance of profitability, demonstration, etc.) and take into account whether PEV technology would be well suited for a GIB intervention.

Notes

1. By taking on projects that may be small yet complex, Australia's CEFC recognises that extra time and resources will likely be needed. The CEFC considers that these transactions create valuable public policy benefits (e.g. demonstration effects) which can lower the level of acceptable financial return (CEFC, 2013).
2. See OECD (2015a) and Kaminker et al. (2013) for further discussion of the role of securitisation in financing LCR infrastructure.
3. NY Green Bank's mission is "to accelerate clean energy deployment in New York State by working in partnership with the private sector to transform financing markets" (NY Green Bank, 2015b) and the Connecticut Green Bank has a mission "to support the governor's and legislature's energy strategy to achieve cleaner, cheaper and more reliable sources of energy while creating jobs and supporting local economic development" (Connecticut Green Bank, 2015a).

References

Australian Treasury (2015), *Portfolio Budget Statement 2015-16: Clean Energy Finance Corporation*, Commonwealth of Australia, www.treasury.gov.au/~/media/Treasury/Publications%20and%20Media/Publications/2015/PBS%202015/Downloads/PDF/08_CEFC.ashx.

Caruso, R. and J. Ellis (2013), "Comparing definitions and methods to estimate mobilised climate finance", *OECD/IEA Climate Change Expert Group Papers*, No. 2013(2), OECD Publishing, Paris, http://dx.doi.org/10.1787/5k44wj0s6fq2-en.

CEFC (2016), "CEFC welcomes creation of Clean Energy Innovation Fund", Clean Energy Finance Corporation, Sydney, Australia, www.cleanenergyfinancecorp.com.au/media/releases-and-announcements/files/cefc-welcomes-creation-of-clean-energy-innovation-fund.aspx.

CEFC (2015a), "CEFC comment on the 2015 Investment Mandate", Clean Energy Finance Corporation, Sydney, Australia, www.cleanenergyfinancecorp.com.au/media/107304/cefc_chairs_response_to_treasurer_and_minister_for_finance_re_2015_cefc_investment_mandate.pdf.

CEFC (2015b), "CEFC has helped accelerate $3.5b in total investment towards a competitive clean energy economy", press release, 15 July, Clean Energy Finance Corporation, Sydney, Australia, www.cleanenergyfinancecorp.com.au/media/releases-and-announcements/files/cefc-has-helped-accelerate-$35b-in-total-investment-towards-a-competitive-clean-energy-economy.aspx.

CEFC (2015c), "Annual report 2014-2015", Clean Energy Finance Corporation, Sydney, Australia, http://annualreport2015.cleanenergyfinancecorp.com.au.

CEFC (2015d) "Transport part of CEFC focus on investment opportunities', *Highway Engineering Australia*, Vol. 47, No. 3, Oct/Nov, Editorial and Publishing Consultants Pty Ltd, http://issuu.com/epcmediagroup/docs/h473_octnov20105_issuu?e=17018673/31293110.

CEFC (2014a), "Annual report 2013-2014", Clean Energy Finance Corporation, Sydney, Australia, www.cleanenergyfinancecorp.com.au/reports/annual-reports/files/annual-report-2013-14.aspx.

CEFC (2014b), "Appendix 2 – Clean energy technologies guidelines", Clean Energy Finance Corporation, Sydney, Australia, www.cleanenergyfinancecorp.com.au/what-we-do/investment-policies/appendix-2-%E2%80%93-clean-energy-technologies-guidelines.aspx.

CEFC (2014c), "CEFC finance for Bindaree Beef biogas and rendering upgrade", Fact Sheet, Clean Energy Finance Corporation, Sydney, Australia, July, www.cleanenergyfinancecorp.com.au/media/76497/cefc-pdf-factsheet-bindaree_lr.pdf.

CEFC (2014d), "CEFC finances deep retrofits to unlock value in commercial property", press release, 28 March, www.cleanenergyfinancecorp.com.au/media/releases-and-announcements/files/cefc-finances-deep-retrofits-to-unlock-value-in-commercial-property.aspx.

CEFC (2013), "Annual report 2012-13", Clean Energy Finance Corporation, Sydney, Australia, www.cleanenergyfinancecorp.com.au/reports/annual-reports/files/annual-report-2012-13.aspx.

CEFC (n.d.), "FAQ", Clean Energy Finance Corporation website, www.cleanenergyfinancecorp.com.au/submit-a-proposal/faqs.aspx#2288 (accessed 20 August 2015).

CEFIA (2014), "Clean Energy Finance and Investment Authority Board of Directors minutes – Regular meeting", 25 April, www.ctcleanenergy.com/Portals/0/board-materials/3_CEFIA__Meeting%20Minutes_042514.pdf.

Cochran, I. et al. (2014), "Public financial institutions and the low-carbon transition: Five case studies on low-carbon infrastructure and project investment", *OECD Environment Working Papers*, No. 72, OECD Publishing, Paris, http://dx.doi.org/10.1787/5jxt3rhpgn9t-en.

Commonwealth of Australia (2015) *Clean Energy Finance Corporation (Investment Mandate) Directions 2015 (No.2)*, ComLaw – Attorney-General's Department, www.comlaw.gov.au/Details/F2015L02114.

Connecticut Green Bank (2015a), "Innovating, educating and activating to accelerate clean energy: 2014 annual report", Connecticut Green Bank, Stamford, Connecticut, www.ctgreenbank.com/wp-content/uploads/2015/12/AnnualReport_FINAL_5.4.15-SinglePages.pdf.

Connecticut Green Bank (2015b), "C-PACE marks successful first two years as CT property owners take advantage of program to finance money-saving energy improvements", press release, Connecticut Green Bank, Stamford, Connecticut.

Connecticut Green Bank (2014), "CEFIA announces sale of commercial property assessed clean energy benefit assessment liens", press release, 19 May, Connecticut Green Bank, Stamford, Connecticut, www.ctcleanenergy.com/NewsEvents/PressRoom/tabid/118/ctl/ViewItem/mid/1364/ItemId/292/Default.aspx?SkinSrc=/Portals/_default/Skins/subpages/subpage_level0.

Connecticut Green Bank (2013), "Connecticut's Green Bank, energizing clean energy finance: 2013 annual report", Connecticut Green Bank, Stamford, Connecticut.

ERB (2015), "New Jersey Energy Resilience Bank Grant and Loan Financing Program Guide", revised 15 October, New Jersey Energy Resilience Bank, www.njeda.com/pdfs/ERB/ERB_ProgramGuide_10_15_15.aspx.

ERB (2014), "New Jersey Energy Resilience Bank Grant and Loan Financing Program Guide", New Jersey Energy Resilience Bank, www.state.nj.us/bpu/pdf/erb/FINAL%20DRAFT%20-%20ERB%20Program%20Guide%208%2020%2014.pdf.

European Commission (2012), "State aid: Commission clears creation of UK Green Investment Bank", press release, 17 October, europa.eu/rapid/press-release_IP-12-1110_en.htm.

Frades, M., J. Peace and S. Dougherty (2014), "The role of clean energy banks in increasing private investment in electric vehicle charging infrastructure", Center for Climate and Energy Solutions, November, www.c2es.org/docUploads/cebs-and-ev-charging-december-2014.pdf.

Gray, G. (2013), Australian Renewable Energy Agency Determination No. 1 of 2013, Australian Renewable Energy Agency Act 2011, Commonwealth of Australia, Ministry of Resources and Energy, Canberra.

IEA (2015), *Medium-term Renewable Energy Market Report 2015*, OECD/IEA Publishing, Paris, http://dx.doi.org/10.1787/renewmar-2015-en.

IEA (2014), "CO_2 emissions from fuel combustion 2014", OECD/IEA Publishing, Paris, www.iea.org/media/freepublications/stats/CO2_Emissions_From_Fuel_Combustion_Highlights_2014.XLS.

IEA (2012), *Energy Technology Perspectives 2012: Pathways to a Clean Energy System*, OECD Publishing, Paris, http://dx.doi.org/10.1787/energy_tech-2012-en.

Kader, B.A. (2015), "Abu Dhabi's carbon capture project on track", Gulfnews.com, 16 February, http://gulfnews.com/business/oil-gas/abu-dhabi-s-carbon-capture-project-on-track-1.1458076.

Kaminker, C. et al. (2013), "Institutional investors and green infrastructure investments: Selected case studies", *OECD Working Papers on Finance, Insurance and Private Pensions*, No. 35, OECD Publishing, Paris, http://dx.doi.org/10.1787/5k3xr8k6jb0n-en.

Kenning, T. (2015), "Buffett project's record low cost part of pricing 'trend', says First Solar", 9 July, PVTech, www.pv-tech.org/news/buffett_projects_record_low_cost_is_part_of_pricing_trend_says_first_solar.

Lazard (2014), "Lazard's levelized cost of energy analysis –Version 8.0, September", http://docplayer.net/30209-Lazard-s-levelized-cost-of-energy-analysis-version-8-0-september-2014.html.

Masdar (2013), "Advancing the clean energy future", corporate brochure, January, www.masdar.ae/assets/downloads/content/226/masdar_corporate_brochure.pdf.

Meza, E. (2015), "US: World's highest prices for Chinese modules", *PV Magazine*, 2 March, www.pv-magazine.com/news/details/beitrag/us--worlds-highest-prices-for-chinese-modules-_100018400/#axzz3Y1g4m2f6.

NREL (2013), "Solar securitization and public capital finance", National Renewable Energy Laboratory, US Department of Energy, Office of Renewable Energy, https://financere.nrel.gov/finance/solar_securitization_public_capital_finance.

NY Green Bank (2015a), "Governor Cuomo announces three NY Green Bank transactions to improve access to clean energy and reduce greenhouse gas emissions", www.governor.ny.gov/news/governor-cuomo-announces-three-ny-green-bank-transactions-improve-access-clean-energy-and.

NY Green Bank (2015b), *Business Plan*, Case 13-M-0412, http://documents.dps.ny.gov/public/Common/ViewDoc.aspx?DocRefId={A112694F-1507-421E-AF8D-DB5D03AD707E}.

NY Green Bank (2014a), *Business Plan*, Case 13-M-0412, http://documents.dps.ny.gov/public/Common/ViewDoc.aspx?DocRefId=%7B3BCF6C87-33FB-49FA-B264-8669055DD6E5%7D.

NY Green Bank (2014b), "RFP 1 clean energy financing arrangements", http://greenbank.ny.gov/rfp1.aspx.

NY Green Bank (2013), "Introducing the New York Green Bank", presentation at the New York State (NYS) Service Commission Technical Conference, 15 October, www3.dps.ny.gov/W/PSCWeb.nsf/96f0fec0b45a3c6485257688006a701a/b6a97b434191050d85257be90047f6af/$FILE/Introducing%20NY%20Green%20Bank.pdf.

New York Public Service Commission (2013), "Petition of the New York State Energy Research and Development Authority to provide initial capitalization for the New York Green Bank", Case 13-M, State of New York Public Service Commission, 9 September, http://documents.dps.ny.gov/public/Common/ViewDoc.aspx?DocRefId=%7BC3FE36DC-5044-4021-8818-6538AAB549B8%7D.

OECD (2015a), *Mapping Channels to Mobilise Institutional Investment in Sustainable Energy*, Green Finance and Investment, OECD Publishing, Paris, http://dx.doi.org/10.1787/9789264224582-en.

OECD (2015b), *Overcoming Barriers to International Investment in Clean Energy*, Green Finance and Investment, OECD Publishing, Paris, http://dx.doi.org/10.1787/9789264227064-en.

Public Utilities Commission of Nevada (2015), "Electronic filing in the matter of the application of Nevada Power Company", submitted 1 July, http://pucweb1.state.nv.us/PDF/AxImages/DOCKETS_2015_THRU_PRESENT/2015-7/3615.pdf (accessed 4 November 2015).

Shahan, Z. (2014), "13 charts on solar panel cost & growth trends", *Cleantechnica*, 4 September, http://cleantechnica.com/2014/09/04/solar-panel-cost-trends-10-charts.

Technology Fund (n.d.), "Target group", Technology Fund website, www.technologyfund.ch/loan-guarantees/target-group (accessed 20 August 2015).

UK Green Investment Bank (2015a), "Annual report 2015", UK Green Investment Bank, Edinburgh, www.greeninvestmentbank.com/about-us/2015-annual-review.

UK Green Investment Bank (2015b), "Smarter, greener cities: Ten ways to modernise and improve UK urban infrastructure", UK Green Investment Bank, Edinburgh, www.greeninvestmentbank.com/media/44748/gib-smarter-greener-cities-report-final.pdf.

UK Green Investment Bank (2014), "Low energy streetlighting: Making the switch", UK Green Investment Bank, Edinburgh, February, www.greeninvestmentbank.com/media/5243/gib-market-report-low-energy-streetlighting-feb-2014-final.pdf.

UK Green Investment Bank (2013a), "About us", UK Green Investment Bank website, www.greeninvestmentbank.com/who-we-are/default.html (accessed 15 January 2016).

UK Green Investment Bank (2013b), "Annual report 2013", UK Green Investment Bank, Edinburgh, www.gov.uk/government/uploads/system/uploads/attachment_data/file/336552/green-investment-bank-annual-report-2013.pdf.

Vivid Economics in association with McKinsey & Co (2011), "The economics of the Green Investment Bank: Costs and benefits, rationale and value for money", report prepared for the Department for Business, Innovation & Skills, October, www.gov.uk/government/uploads/system/uploads/attachment_data/file/31741/12-554-economics-of-the-green-investment-bank.pdf.

Chapter 3.

Types of green investment bank interventions and co-investors

This chapter reviews the types of investments that green investment banks undertake, the types of instruments and funds they use to invest and the co-investors they attract. It examines the range of de-risking approaches used by green investment banks and their innovative approaches to reduce the high transaction costs often associated with low-carbon and climate-resilient infrastructure investments. The chapter closes with a discussion of the types of private investors green investment banks collaborate with or seek to attract.

Key takeaways

- Green investment banks directly finance investment in low-carbon and climate-resilient infrastructure using a range of instruments and funds, including senior and subordinate loans, bond-based financing and equity.
- Green investment banks also provide risk-mitigating credit enhancements such as loan loss reserves and guarantees to reduce risks for private sector lenders.
- Green investment banks can encourage the adoption of repayment mechanisms such as on-bill finance, which facilitates repayment through existing utility bills and reduces default risk for lenders.
- Transaction enablers such as warehousing and securitisation increase the flow of institutional capital by bundling small-scale projects to achieve scale and reduce transaction costs.
- Many types of private investors have co-invested with green investment banks, including local lenders, investment banks, institutional investors and retail investors. Green investment banks that are mandated to promote smaller scale investment are likely to collaborate with local lenders and energy service companies while entities that seek to support large-scale projects often target institutional investors and investment banks.

Green investment bank interventions

In order to attract private investment for low-carbon and climate-resilient (LCR) infrastructure, a green investment bank (GIB) identifies the primary obstacles perceived by investors and adapts its interventions accordingly. Generally these interventions aim to reduce the risks and transaction costs for such investments, improve returns and better align the risk-return profile of LCR projects with the requirements of different types of investors. The precise intervention will depend on the flexibility a GIB has been given in its mandate. For example, NY Green Bank has broad authority to use a range of investment instruments, funds and structures, although it is required to update its business plan and strategic direction annually based on lessons learnt and market conditions. This broad authority presumes a high level of expertise and gives the institution flexibility to pursue a wider range of projects and attract additional types of co-investors. Malaysia's Green Technology Financing Scheme (GTFS), in contrast, is chartered to provide a single form of financing. This "standard-offer" approach simplifies deal consideration and structuring expertise in a given area, but it may limit the range of projects to which GreenTech Malaysia can attract private investment.

Given that GIBs invest using public capital, their involvement in a commercial transaction can provide private investors with additional certainty. GIB investments may take several forms. For example, a GIB may take an equity ownership stake in a company or it may lend money to a project through a loan with a specified repayment structure. This financial investment is often paired with a transaction-enabling component to mitigate risks and reduce transaction costs. Risk mitigants that support private investors require a financial commitment by the GIB (such as a loan guarantee or loan loss reserve). Other transaction enablers, like on-bill financing or warehousing, facilitate investment but do not in themselves require additional GIB capital commitment.

The following sections describe GIB instruments and funds, along with types of transaction enablers and risk mitigants. The OECD publication *Mapping Channels to Mobilise Institutional Investment in Sustainable Energy* defines the full range of instruments, funds, risk mitigants and transaction enablers that comprise the key elements of renewable energy deals, and provides examples from a database of nearly 70 transactions (OECD, 2015).

Instruments and funds

Investment instruments and funds used by GIBs include the following:

- **Loans:** Loans may be provided for projects or companies and can include both senior and junior debt. Loans are the most common GIB investment instrument and are provided by every operational GIB with the exception of Masdar, the Swiss Technology Fund and the Japanese Green Fund. Since project-specific financing typically requires a bank loan in addition to the project owner's equity investment, projects that fail to secure private loan financing often do not reach the construction phase. GIB loans can fill this financing gap, as GIBs can set loan terms to match the revenue or energy savings stream of a given project, improving the likelihood of repayment.

- **Equity:** GIBs also make equity investments in both projects and companies. In order to draw in private sector equity, GIBs generally do not take majority stakes. In addition to providing a source of capital, GIB equity investments may also indirectly act as a risk mitigant for private investors. Equity investments carry a lower priority for repayment if a project or company faces a default or bankruptcy. GIB equity investments thereby give private debt investors participating in a project greater confidence that they will recoup their investments, as they have a higher priority for repayment and can also assume that the GIB seeks to recover its equity investment. The UK Green Investment Bank, the Connecticut Green Bank, the Japanese Green Fund and Masdar make equity investments.

- **Mezzanine capital:** Mezzanine capital is a type of hybrid financing that begins as debt and gives the lender the right to convert it to an ownership or equity interest in the company if the loan is not paid back in time or in full. Some GIBs such as those in Australia, Japan, the United Kingdom, and in Connecticut and New York in the United States specifically mention mezzanine capital as a permitted investment instrument. In 2014, the UK Green Investment Bank made a GBP 16.9 million mezzanine debt investment in a GBP 110 million waste wood (renewable) combined heat and power (CHP) plant, alongside multiple private investors making senior, mezzanine and equity investments (UK Green Investment Bank, 2014a).

- **Investment funds:** Investing in existing funds can be attractive for GIBs that wish to support smaller projects, such as individual energy efficiency projects. For example, the Connecticut Green Bank made equity investments in a solar lease fund, which is used to finance many small distributed solar generation projects. GIBs can also set up their own debt or equity investment funds. In April 2015, the UK Green Investment Bank reached a first close of GBP 463 million for a fund to support offshore wind development (the Operating Offshore Wind Fund), for which it intends to provide 20% of capital when it reaches its full size of GBP 1 billion. It reached a second close of GBP 818 million in October 2015, securing investment from UK-based pension funds and international institutional investors, including a large sovereign wealth fund. New investments allow project developers to sell their stakes and finance new projects (UK Green Investment Bank, 2015). Other examples of fund investment include Masdar Capital, a division of Masdar, which has established funds that have attracted private investors as limited partners.[1]

- **Bonds:** Some GIBs can issue bonds through a public or private sale in order to capitalise the GIB itself or to recapitalise a loan warehouse. By issuing bonds, GIBs can draw large amounts of private institutional capital to LCR infrastructure investment, and depending on legal authority, a GIB may be able to issue government-backed bonds. This facilitates lower interest rates, enabling the GIB to lend the funds at a lower cost of capital.

  In addition to using bond issuances as a tool for initial capitalisation, the UK Green Investment Bank has expressed interest in issuing bonds to refinance its investments in green infrastructure and other green investments. The Connecticut Green Bank securitised its commercial and industrial energy efficiency and renewable energy loans (secured by a lien on the property as explained later in Chapter 4), resulting in a sale of senior bonds to an institutional investor with the GIB retaining the junior bonds. This transaction is discussed further in Chapter 5.

- **Structured notes:** Some GIBs, notably the Connecticut Green Bank, have issued "structured notes" backed by pools of collateral. A structured note is a debt obligation that is structured to deliver the risk-return performance of another type of investment by means of investing in a derivative for that type of investment. In Connecticut's case, a "bankruptcy remote" special purpose entity was established to hold a pool of solar loans against which structured notes were issued and sold to accredited investors via two crowdfunding platforms, which allowed a range of investors to invest against a pool of assets or even in individual projects.

- **Grants:** The Connecticut Green Bank was tasked in its authorising statute to manage and wind-down over time the state's residential roof-top solar grant programme (State of Connecticut, 2011).

Transaction enablers and risk mitigants

To attract private investment, GIBs can make private sector lending less costly, reduce liquidity risk, make new markets more accessible, or reduce the risk of repayment or default. GIBs often pair their investment instruments and funds with transaction-enabling structures or risk mitigants to attract private investment.

These enabling and risk-mitigating approaches presume that private investors are willing to invest in GIB target markets on near-commercial terms, provided they receive a nudge from the public sector to facilitate an investment. In cases where a project is particularly risky, or where the risk is difficult to calculate due to the innovative nature of the project or where local markets are insufficiently developed, a GIB may conclude that grants or grant-like methods are needed to facilitate greater private investment.

Structuring and product-design methods used by GIBs to enable transactions include:

- **Warehousing:** Warehousing is an aggregation technique used to reduce transaction costs and facilitate investment. Small projects are bundled together to reach a scale where they become attractive for on-sale to large investors or for securitisation through bond issuances. Aggregation techniques such as loan warehousing can reduce transaction costs and facilitate investment in bundled small-scale projects, thereby helping them reach commercial scale. For example, in addition to bundling and securitising commercial and industrial loans, the Connecticut Green Bank has combined solar leases from a large number of small residential projects to attract private companies and new investors through its Solar Lease II programme. By reducing transaction costs and increasing scale, the Connecticut Green Bank's warehousing attracted new investors as well as providers of insurance that can facilitate investments. For example, Mosaic, a solar finance "crowdsourcing" company, will "crowdsource" USD 5 million for a pool of loans (Business Wire, 2014) and a private insurance company created a new product to provide insurance and warranties for solar leases. Aggregation techniques and bundling of small-scale projects could be instrumental to increasing potential projects to a commercially attractive size.

- **Securitisation:** Securitisation is a technique whereby non-traded or small-scale assets, such as cash flows from solar leases or power-purchase agreements, are transformed into a standardised, tradable asset. By warehousing or aggregating smaller transactions, GIBs can take a pool of loans or leases and securitise it by issuing bonds to be repaid from the proceeds of the loan pool, or by providing bond-like returns or dividends on capital investments in the securitised pool of loans. NY Green Bank has participated in a national energy efficiency warehousing and securitisation platform (WHEEL, described in Chapter 4) that provides institutional investors access to residential energy efficiency project investments (NY Green Bank, 2015).

- **On-bill financing:** On-bill financing allows borrowers to repay renewable energy or energy efficiency loans through an additional charge on their existing utility bill. This facilitates customer repayment and reduces the risk of default for an investor. Customers place a high priority on maintaining electricity service and will pay their electricity bill at a high rate. Early data indicate that default rates on on-bill financing are low, ranging between 0% and 3% (LBNL, 2014). Australia's CEFC partnered with an Australian energy retailer to provide on-bill financing for businesses that undertake energy efficiency upgrades or install solar photovoltaic (PV) panels; financing is available for up to seven years for projects between AUD 50 000 and AUD 1 million. Hawaii's Green Energy Market Securitization (GEMS) programme, which will provide financing for residential roof-top solar, also included on-bill repayment in its structure as a strategy to induce repayment and reduce potential default rates (Strand and Seligman, 2013). The Connecticut Green Bank is developing an "open source" on-bill repayment programme whereby a diverse group of lenders and capital providers (such as banks, credit unions and solar leasing companies) can provide financing for solar PV loans and leases, and will be able to collect loan and lease payments through a utility bill charge.

- **Financing through tax payments:** The Commercial Property-Assessed Clean Energy (C-PACE) programme supported by the Connecticut Green Bank provides upfront financing for energy efficiency and renewable energy upgrades that are repaid through property taxes over time. By using a tax lien, C-PACE provides a long-term and secure product for private investors as these loans are repaid in a steady stream alongside tax payments. Repayment obligations are also transferred to the next owner if the property is sold. As discussed further in Chapter 4, the Connecticut Green Bank has sold these loans to private capital providers, offering stable returns and freeing up the bank's capital to make additional loans. Australia's CEFC uses a similar technique in its environmental upgrade agreements by funding building energy efficiency improvements that are repaid through a local council charge on the land. Borrowers face far greater consequences if the repayment is part of their tax liability and typically have lower default rates compared to other forms of debt. Loans tied to property taxes are more secure than other loans because the property acts as collateral, and so loans are perceived as less risky by creditors and, as a result, may be awarded better terms, such as 15-20-year repayment periods.

- **Leasing:** Leasing can provide an attractive alternative to purchasing residential, commercial or industrial renewable energy or energy efficiency technologies. For example, leasing solar PV panels for rooftop applications is the leading entrance point to the residential solar sector in the United States (Munsell, 2014). Customers enjoy the benefits of self-generated renewable energy, at a lower cost than the utility, but without the burdens of ownership, like maintenance. However, many leasing options have strict credit limits, which reduce the pool of eligible individuals to zero in some regions. In this context, GIBs can support solar PV leasing by co-investing in lease funds with private debt or equity partners.

Credit-enhancing risk mitigants reduce the risk that a project or investment will not deliver its expected level of return and can take the form of loan loss reserve funds or loan guarantees. Subordinate debt or equity can also indirectly serve as a credit enhancement. GIBs provide the following:

- **Loan loss reserve funds:** Loan loss reserves set aside capital to cover potential losses and help to reduce repayment risk. If a borrower (such as a purchaser of a solar PV installation) defaults, the lender (such as an institutional investor) is repaid using the reserve fund. GIBs may provide a percentage of loan loss coverage for lenders. As part of its Smart-E Loan Program, the Connecticut Green Bank offers distinct residential energy efficiency and renewable energy financing products with corresponding loan loss reserve levels. Every time a lender underwrites a loan, the Connecticut Green Bank reserves a percentage of the loan principal (7.5-15%) for the lender in the event of a default (Energize CT, 2013). In the Connecticut model, to promote sound lending practices and share risks, the lender assumes the "first loss" (1.5%) on its portfolio before it can access the reserve. NY Green Bank has also listed loan loss reserves as a viable credit enhancement structure to be used to support investments (NY Green Bank, n.d.).
- **Guarantees:** Guarantees are a credit enhancement tool used to mitigate perceived or actual risks to improve the attractiveness of investments, often debt instruments. By providing a loan guarantee, a guarantor (such as a GIB) agrees to pay a lender a portion of the loan if a borrower cannot repay. For example, GreenTech Malaysia provides guarantees to encourage private banks to finance green projects. Its Green Technology Financing Scheme (GTFS) assesses applications for "green project certificates" and provides certificates to eligible

companies. These certified companies can then seek loans from participating private lenders. In order to improve lending approval rates and reduce risk, the GTFS guarantees repayment of 60% of the financing provided by private lenders to certified companies in the event of loan default. Australia's CEFC is authorised to provide loan guarantees, but it seeks to avoid providing them and limits guarantees to 5% of the total CEFC portfolio (CEFC, n.d.).

- **Subordination:** A GIB can increase the likelihood of repayment for private investors by making subordinate debt or equity investments in a project alongside private investors. In the event of default, any remaining value or cash from the project is paid out to investors in the order of seniority, with senior investors repaid before subordinate investors. In addition to taking an equity position, the UK Green Investment Bank made a commercial GBP 16.9 million mezzanine debt investment in a CHP plant in 2014. This subordinate debt position supported GBP 42.5 million in senior private debt investment (UK Green Investment Bank, 2014a). In 2013, the Connecticut Green Bank took both an equity and subordinated debt position in a residential solar lease fund it established with private lenders (Connecticut Green Bank, 2013).

Green investment bank co-investors

A GIB's mandate and strategy will determine the targeted type of private investment. These strategies can be characterised as "wholesale" or "retail". A wholesale strategy seeks to attract relatively large amounts of private capital to combine with public capital to use to on-lend or invest in funds. A retail approach, in contrast, involves delivery of funds to the project developer or individual. Wholesale lending can move large volumes of investment while retail lending can be useful for jump-starting activity in new markets. Under either scenario, a GIB may help bring projects to a broader set of potential investors through bond issuances, securitisation or private placement.

Local banks

Local banks can play a valuable role in issuing individual loans to residential or commercial borrowers. An individual home or business owner interested in improving the efficiency of their building or in installing distributed renewable energy generation might seek a specialised financing firm (e.g. SolarCity in the case of solar PV) or directly approach their local lending institution. Drawing local banks into the LCR space can help GIBs to grow their target markets, as many potential borrowers have already established banking relationships with their lenders. GIBs can support and provide capital through on-lending to local banks to ensure they are able to offer attractive loans.

Some local banks already offer financing products specifically designed to serve renewable energy and efficiency borrowers. However, the majority of lenders are unaware of, or averse to, investing in this growing market due to perceived repayment risk or limits on unsecured lending. Much of this perception is due to short technology track records or uncertainty regarding technology performance, especially for energy efficiency investment. In addition, local lenders often do not account for the expected financial savings of an energy investment during the underwriting process and instead focus on the pure credit rating of the borrower. Based on this approach, banks overestimate the repayment risk and as a result limit the pool of acceptable borrowers, require high interest rates or provide short loan tenors.

These financial drawbacks are often compounded by limited marketing and consumer engagement by local banks. The most common way for a potential borrower to learn about an available lending product is through a contractor or service provider that assesses or installs the renewable or efficient technology. If the local lender has not informed contractors about available financing, the information may never reach customers. Even in cases where contractors are able to direct customers to a local bank, it is often the customer's responsibility to co-ordinate the activity of the bank and the contractor. This may prove to be a significant barrier to adoption for customers that are unsure of the merits of such an investment.

GIBs can crowd in more private investment from local banks either by directly partnering on retail lending or by purchasing lenders' loans to provide liquidity. Direct partnership with local lenders could involve co-lending to borrowers or offering lenders a credit enhancement to incentivise more lending activity. Connecticut's Smart-E Loan Program takes the latter approach, offering a loan loss reserve to local banks that provide energy efficiency or renewable loans to residential customers. Connecticut also has co-lended to several projects, including a 15 MW grid-tied fuel cell, a 5 MW grid-tied wind facility and a 2 MW anaerobic digestion facility, in each case using subordinated debt at interest rates ranging from concessional to market rate. Malaysia's experience with Islamic banks is discussed in Box 3.1.

Box 3.1. **Islamic finance and Malaysia's Green Technology Financing Scheme**

Malaysia is considered a pioneer in Islamic finance with a strong and growing Islamic banking sector. Islamic banking, also known as *sharia*-compliant banking, is consistent with principals of the *Sharia* (Islamic rulings). The *Sharia* prohibits payment or acceptance of interest charges for lending and also prohibits supporting activities that are considered to be sinful, such as alcohol consumption or gambling.

Malaysia's Green Technology Financing Scheme (GTFS) believes the principles of Islamic finance are well aligned with green and socially responsible investing and is actively working with Islamic banks to attract private capital. GreenTech Malaysia's CEO, Ahmad Hadri Haris, said, "We have identified Islamic banking as a platform, as it is based on the promotion of value and good practices." All Islamic banks are eligible to become participating financing institutions under the GTFS' loan guarantee programme and Islamic financing accounts for 40% of all funds granted under the GTFS (Bank Negara Malaysia, 2014).

Sources: Bank Negara Malaysia (2014), "Islamic finance: Ready to finance a greener world", www.mifc.com/?ch=28&pg=72&ac=88&bb=uploadpdf; Bernama (2014), "Greentech Malaysia to approve more funds", 4 February, www.ibfim.com/img/media-centre/media-coverage/media-coverage-20140204.pdf.

GIBs could also directly on-lend to local banks. Multilateral development banks have significant experience with this strategy. For example, the European Investment Bank lends to local banks which in turn on-lend to smaller borrowers. Sub-national GIBs are particularly well suited for this kind of activity due to their knowledge of local banks and market conditions. An alternative strategy would be for a GIB to launch a fund that purchases renewable energy loans from a local lender to remove the origination burden from the GIB and to encourage local banks to become familiar with renewable energy loans. This addresses a concern held by local banks that renewable energy loans are illiquid assets that will remain on the bank's balance sheets with no way to recapitalise the pool of loans.

Investment banks

Investment banks are increasingly active in LCR investment and can be valuable private investment partners for GIBs. For example, the investment banking and financial services corporation Citi announced a ten-year, USD 100 billion commitment to finance sustainable growth in 2015 (Citi, 2015). In 2014, Bank of America launched the Catalytic Finance Initiative, with a goal to stimulate at least USD 10 billion in renewable energy projects (Bank of America, 2014). Since then, the Initiative has evolved into a consortium that includes other financial organisations with their own capital commitments. In April 2016, the consortium committed USD 8 billion towards "high-impact sustainable investments" (IFC, 2016). Investment banks are capable of channelling large amounts of invested capital. GIBs can work with investment banks to identify investment opportunities that are attractive for both parties.

In addition to direct investment, investment banks can help GIBs mobilise large pools of capital through securitisation, especially as securitisations increase in scale and when expertise in asset-backed debt securitisation is required. For an investment bank to underwrite a security issuance, it must be comfortable taking on the debt and risk associated with the underlying investment. Investment banks can be hesitant to be the first mover on a new type of transaction, and GIBs can support securitisations by standardising the underlying loans and credit requirements and by warehousing smaller loans into a large portfolio.

Institutional investors

Institutional investors are an important potential source of alternative capital for domestic LCR infrastructure investment. They include insurance companies, pension funds, investment funds, public pension reserve funds, foundations, endowments and other forms of institutional savings. In OECD countries alone, these investors held USD 93 trillion[2] of assets in 2013 (OECD, 2014a; 2014b). Despite their significant size, institutional investors' asset allocation to direct infrastructure investments in general remains small, less than 1% for large OECD pension funds, and the "green" investment component remains even more limited. This investment is constrained for a variety of reasons, including regulatory and policy uncertainty, a lack of suitable financing vehicles, investor inexperience with direct investing in new technologies and asset classes, as well as market and government failures (OECD, 2015). Institutional investors often seek long-term and low-risk investments, and allocate significant amounts of capital domestically. Institutional investors are also generally uncomfortable taking on construction risks or being the first movers into a new market.

Some GIBs are looking to engage institutional investors as the deepest and most accessible pool of global capital. In an initial study on the prospect of creating a green investment bank in the United Kingdom, Ernst & Young (2010) emphasised the importance of creating investment opportunities attractive to institutional investors as well as actively interacting with these investors during the design phase of the institution. Ernst & Young (2010) recommended that the UK Green Investment Bank "act as a bridge between institutional capital and ultimate investments" and "should be strategically structured to appeal to the widest and deepest sources of capital as possible". The Chairman of the Board of the UK Green Investment Bank also highlighted the importance of engaging institutional investors in an address at a National Association of Pension Funds Conference (UK Green Investment Bank, 2014b). The initial concept for NY Green Bank also envisioned direct engagement with institutional investors. The original petition to provide the bank's initial capitalisation proposed that "the Bank could execute a debt securitisation, through which investors interested in holding long-term debt, such as pension funds, could invest in longer term securities, while those banks preferring shorter loan terms would be able to exit their investments earlier" (New York Public Service Commission, 2013). To date, GIBs have attracted institutional investment using a variety of instruments and funds, risk mitigants and transaction enablers, which are outlined below.

- Cornerstone stake: A cornerstone investment refers to a large investment in a company or fund that occurs early in the investment process so as to play a demonstration role and attract other investors. GIBs have taken cornerstone stakes to attract pension and insurance capital.
- Co-investing: GIBs can co-invest by providing debt or equity for a project or company. This investment can support new investment or help to recycle capital through refinancing. For example, the UK Green Investment Bank participated in a loan consortium to refinance a stake in the Walney Offshore wind farm owned by Ampere Equity Fund and Dutch pension fund PGGM.
- Issuing green bonds: Bonds are an asset class favoured by institutional investors. NYSERDA, the parent agency of NY Green Bank, issued a USD 26 million bond in 2013 to securitise a portfolio of residential and small commercial sector energy efficiency loans. The bond used an innovative structure and federal tax benefits to secure an AAA rating (CE+BFI, 2013).
- Developing funds: The UK Green Investment Bank created a fund that will invest in multiple offshore wind projects. The fund is designed to appeal to institutional investors that may seek exposure to assets such as offshore wind but would be unlikely to risk investing in a single project (Shankleman, 2014).
- Selling loan portfolios: The Connecticut Green Bank secured USD 100 million from a Real Estate Investment Trust (REIT) in December 2015 for its C-PACE programme. The REIT has committed to fund a portfolio of PACE financings being originated by the bank for energy updates in commercial buildings. The REIT can be considered institutional money as it is publicly traded and as investment in REITs tends to be dominated by institutional investors (personal communication with Bert Hunter, Connecticut Green Bank, 1 February 2016).
- Loan warehousing to facilitate securitisation: GIBs can structure prospective investment opportunities to have long-term cash flows that will be attractive to institutional investors. The Connecticut Green Bank and NY Green Bank have designed loan-bundling programmes to facilitate securitisation or sell-offs to larger investors. For example, the Connecticut Green Bank warehoused and securitised its commercial energy efficiency PACE loans and sold them through a private placement to Clean Fund, an institutional investor specialised in PACE investments (Lombardi, 2014).

Individual retail investors

In many financial markets, LCR infrastructure investment opportunities for individuals are limited. Renewable energy and energy efficiency projects face barriers to raising capital through public capital markets, and due to limited offerings it is difficult for individuals to buy stock in a project or group of projects, or to purchase shares in LCR-specific funds.

GIBs can directly or indirectly facilitate individual retail investment in LCR infrastructure through partnering with "crowdfunding" investment platforms to pool individual projects. GIBs have already demonstrated the use of these types of partnerships: in 2014, the Connecticut Green Bank sold a portion of a solar loan fund to Solar Mosaic, a solar-specific crowdfunding platform, which in turn has funded those loans through individual, crowdsourced investments (Business Wire, 2014).

GIBs can indirectly facilitate individual retail investment by building the structures needed to link renewable energy investment with public capital markets and create needed scale and consistency. For example, GIBs can pool small loans involving similar technologies and underlying credit risks and can set credit parameters so that only borrowers above a certain credit score are eligible.

Specialised service and financing firms

GIBs can work with specialised renewable energy and energy efficiency financing firms, such as energy service companies (ESCOs)[3] or solar PV lease or power purchase agreement (PPA) providers, to offer customers an integrated energy service and financing solution, subject to these providers' constraints. For instance, ESCOs prefer to serve large commercial or industrial customers that have large facilities and high credit ratings, while solar PV leasing and PPA firms often may only offer financing to residential customers with high credit scores.

A GIB can also provide a credit enhancement to a specialised firm to extend its market reach. In 2014, NY Green Bank announced that it had reached an agreement in principle to provide a credit facility (corporate loan or collection of loans) to Ameresco, a large national ESCO, to be used in partnership with private third-party financing to address underserved segments of commercial and industrial energy efficiency markets (NY Green Bank, 2014).

Notes

1. The Masdar Clean Tech Fund was launched in 2006 and is co-managed by Masdar Capital, Consensus Business Group and Credit Suisse (Masdar, 2012b). This USD 250 million venture capital fund invests in early-stage clean technologies. Masdar's second fund (USD 290 million), the DB Masdar Clean Technology Fund, was developed in conjunction with Deutsche Bank and invests in renewable energy, environmental resources, and energy and material efficiency (Masdar, 2012a).

2. This figure includes assets of pension funds and insurance companies which may be also counted in investment funds.

3. An ESCO acts as a third-party installer and financier for building efficiency upgrades. ESCOs are able to offer building owners energy savings through efficiency with no upfront cost, with financing paid back over time by the ESCO sharing a portion of the energy savings over time. ESCOs may offer equipment financing as part of their services and typically only serve large and credit-rated industrial customers.

References

Bank Negara Malaysia (2014), "Islamic finance: Ready to finance a greener world", www.mifc.com/?ch=28&pg=72&ac=88&bb=uploadpdf.

Bank of America (2014), "Bank of America announces $10 billion catalytic finance initiative to accelerate clean energy investments that reduce carbon emissions", News Release, 23 September, http://newsroom.bankofamerica.com/press-releases/corporate-and-investment-banking-sales-and-trading-treasury-services/bank-america-ann.

Bernama (2014), "Greentech Malaysia to approve more funds", 4 February, www.ibfim.com/img/media-centre/media-coverage/media-coverage-20140204.pdf.

Business Wire (2014), "Sungage Financial, CEFIA, and Mosaic announce $5 million deal to offer new, crowdsourced residential solar loans," Business Wire, 6 February, www.businesswire.com/news/home/20140206005031/en/Sungage-Financial-CEFIA-Mosaic-Announce-5-Million#.VQI3j_nF_NE.

CE+BFI (2013), "Anatomy of a deal," Clean Energy and Bond Finance Initiative, Clean Energy Group and CDFA, September, www.cdfa.net/cdfa/cdfaweb.nsf/ord/cebfi-anatomyofthedeal-nyserda.html/$file/CEBFI%20NYSERDA%20Anatomy%20of%20the%20Deal%20-%20Final.pdf.

CEFC (n.d.), "Investment Strategy", Section 5 of CEFC Investment Policies, Clean Energy Finance Corporation, Sydney, Australia, www.cleanenergyfinancecorp.com.au/what-we-do/investment-policies/investment-strategy.aspx (accessed 12 March 2015).

Citi (2015), "Citi announces $100 billion, 10-year commitment to finance sustainable growth", Citigroup Inc, 18 February, www.citigroup.com/citi/news/2015/150218a.htm.

Connecticut Green Bank (2013), "CT Solar Lease 2 Program", Clean Energy Finance and Investment Authority, 26 June, www.ctcleanenergy.com/Portals/0/board-materials/CEFIA_Due_Diligence_Package_Programmatic_Solar_Lease_2.pdf.

Energize CT (2013), "About smart-e loans", Lender Fact Sheet, Energize Connecticut, 18 March, www.ctcleanenergy.com/Portals/0/board-materials/6a_Smart-E%20Loan_Lender%20Fact%20Sheet.pdf.

Ernst & Young (2010), "Capitalising the green investment bank: Key issues and next steps", Ernst & Young LLP, London, http://e3g.org/docs/capitalisingthegreeninvestmentbank.pdf.

IFC (2016), "Announcement: Catalytic Finance Initiative directs $8 billion in capital for high-impact sustainable projects", http://www.ifc.org/wps/wcm/connect/news_ext_content/ifc_external_corporate_site/news+and+events/news/catalytic-finance-initiative.

LBNL (2014), "Financing energy improvements on utility bills: Market update and key program design considerations for policymakers and administrators", presentation, Lawrence Berkeley National Laboratory, April, www4.eere.energy.gov/seeaction/system/files/documents/publications/chapters/onbill_financing_summary.pdf.

Lombardi, N. (2014), "In a 'watershed' deal, securitization comes to commercial efficiency," Greentech Media, 19 May, www.greentechmedia.com/articles/read/the-first-known-commercial-efficiency-securitization.

Masdar (2012a), "Masdar Clean Tech Fund", Masdar, www.masdar.ae/en/home/detail/masdar-clean-tech-fund (accessed 12 January 2016).

Masdar (2012b), "DB Masdar Clean Tech Fund", Masdar, www.masdar.ae/en/home/detail/db-masdar-clean-tech-fund (accessed 12 January 2016).

Munsell, M. (2014), "Market share for leasing residential solar to peak in 2014", Greentech Media, 23 June, www.greentechmedia.com/articles/read/Market-Share-for-Leasing-Residential-Solar-to-Peak-in-2014 (accessed 13 January 2016).

NY Green Bank (2015), "Expanding availability of residential energy efficiency loans to homeowners throughout New York state", Transaction Profile, September, http://greenbank.ny.gov/-/media/greenbanknew/files/Residential-Energy-Efficiency-October-2015.pdf.

NY Green Bank (2014), "NY Green Bank's initial transactions", NY Green Bank website, http://greenbank.ny.gov/initial-transactions.

NY Green Bank (n.d.), "Product offerings", NY Green Bank website, http://greenbank.ny.gov/Approach/Product-Offerings (accessed 13 January 2016).

New York Public Service Commission (2013), "Petition of the New York State Energy Research and Development Authority to provide initial capitalization for the New York Green Bank", Case 13-M, State of New York Public Service Commission, 9 September, http://documents.dps.ny.gov/public/Common/ViewDoc.aspx?DocRefId=%7BC3FE36DC-5044-4021-8818-6538AAB549B8%7D.

OECD (2015), *Mapping Channels to Mobilise Institutional Investment in Sustainable Energy*, Green Finance and Investment, OECD Publishing, Paris, http://dx.doi.org/10.1787/9789264224582-en.

OECD (2014a), "Balance sheet and income", *OECD Insurance Statistics* (database), http://dx.doi.org/10.1787/ins-data-en.

OECD (2014b), "Pension statistics", *OECD Pensions Statistics* (database), http://dx.doi.org/10.1787/pension-data-en.

Shankleman, J. (2014), "Green Investment Bank mulls pension fund boost for offshore wind industry", Business Green, 1 April, www.businessgreen.com/bg/analysis/2337219/green-investment-bank-mulls-pension-fund-boost-for-offshore-wind-industry.

State of Connecticut (2011), Senate Bill No. 1 243, Public Act No. 11-80, 1 July, www.cga.ct.gov/2011/act/pa/pdf/2011PA-00080-R00SB-01243-PA.pdf.

Strand, M. and J. Seligman (2013), "Renewable energy installations accelerate in Hawaii", Chadbourne & Park LLP, Washington, August, www.chadbourne.com/files/Publication/98a55a15-4c0c-468c-a0df-

859432bbbda6/Presentation/PublicationAttachment/f0243bec-8417-4fbf-9ae9-8fac2acdab4f/RenewableEnergyInstallationsInHawaii_Seligman_Aug13.pdf.

UK Green Investment Bank (2015), "Delivering value in the UK's offshore wind market", UK Green Investment Bank Plc, www.greeninvestmentbank.com/funds/offshore-wind-fund.

UK Green Investment Bank (2014a), "£110m funding secured for biggest waste wood renewable energy facility in North West", UK Green Investment Bank, 21 November, www.greeninvestmentbank.com/news-and-insight/2014/110m-funding-secured-for-biggest-waste-wood-renewable-energy-facility-in-the-north-west.

UK Green Investment Bank (2014b), "Lord Smith of Kelvin address to the National Association of Pension Funds", Speech to National Association of Pension Funds Investment Conference, Edinburgh, 6 March, www.greeninvestmentbank.com/news-and-insight/2014/lord-smith-of-kelvin-address-to-the-national-association-of-pension-funds.

Chapter 4.

Green investment banks and energy efficiency

This chapter discusses how green investment banks are working to reduce barriers for private investment in energy efficiency and explores the range of interventions they use to scale up energy efficiency investment. It also describes the investment partnerships that green investment banks pursue in the field of energy efficiency.

Key takeaways

- Energy efficiency represents a massive investment opportunity across multiple sectors, such as industry, buildings and power generation, with a global market valued in the hundreds of billions (USD). Green investment banks are addressing multiple barriers to energy efficiency investment in buildings, including:
  - small average investment size and relatively high transaction costs
  - the need for long-term capital to match the flow of savings
  - difficulty in measuring or underwriting energy savings
  - lack of familiarity among private investors.
- Many of the investments green investment banks mobilise are undertaken in urban areas where 54% of the world's population lived in 2014 and where 66% is projected to live by 2050.
- Green investment banks use a range of credit-enhancement and direct investment mechanisms to deploy public capital and leverage private investment in energy efficiency.
- On-bill financing and linking energy efficiency loan repayment to tax payments through tax liens are two innovative structures that increase the chances of repayment and reduce risks for the lender.
- Green investment banks are developing efficiency-focused funds and providing direct lending and leasing offerings to fill gaps in the efficiency lending marketplace.
- Green investment banks can attract large institutional investors by warehousing smaller energy efficiency loans and then selling those loans at scale through securitisation.

Introduction to energy efficiency investment

The opportunity of increased energy efficiency and key sectors

The International Energy Agency (IEA) describes energy efficiency as the "first fuel" because energy efficiency improvements satisfy more energy demand than any single fossil fuel (IEA, 2014a).[1] Energy efficiency investments are a central part of national greenhouse gas (GHG) emissions mitigation strategies and energy planning as they reduce energy consumption, lower GHG emissions and provide savings from avoided investments in generation capacity and transmission and distribution. They also provide multiple benefits beyond GHG reductions, such as reduced air pollution and improved energy security (Box 4.1).

Significant energy efficiency opportunities exist across all sectors. However, due to the many barriers that limit the uptake of energy efficiency, such as split incentives, information failures and subsidised energy prices, the IEA estimates that two-thirds of "economically viable"[2] energy efficiency potential will remain unrealised (IEA, 2014b). As transport and industry have already made important energy efficiency gains, the sectors with the greatest unrealised potential for energy efficiency are buildings and power generation (IEA, 2014b). This chapter focuses primarily on green investment bank (GIB) activities to facilitate the financing of energy efficiency projects in buildings, many of which are undertaken in cities (Box 4.2).

Box 4.1. The multiple benefits of energy efficiency

Improving energy efficiency can provide a range of benefits to different stakeholders. The IEA study *Capturing the Multiple Benefits of Energy Efficiency* identified 15 distinct benefits of energy efficiency. These include:

- macroeconomic development through energy efficiency investment that can increase employment and economic activity
- reduced strain on public budgets through reduced government expenditures on fuel for heating, cooling and lighting
- improved health and well-being as a result of energy efficiency retrofits and weatherisation programmes that can reduce respiratory, cardiovascular and allergy risks, and stress
- greater industrial productivity through energy efficiency can enhance competitiveness, increase productivity and improve working environments
- improved energy delivery though reduced energy generation, transmission and distribution costs, greater system reliability and less volatility in wholesale markets.

Governments can employ a range of measures and policies to stimulate demand for energy efficiency investments. For example, green investment banks can serve as a key element of a country's (or sub-national jurisdiction's) policy framework for energy efficiency investment. At the international level, there is increasing recognition of the importance of domestic policies to support energy efficiency investment. In October 2015, G20 Energy Ministers welcomed the Voluntary Energy Efficiency Investment Principles for G20 Participating Countries.

Sources: Ryan, L. and N. Campbell (2012), "Spreading the net: The multiple benefits of energy efficiency improvements", *IEA Energy Papers*, No. 2012/08, OECD Publishing, Paris, http://dx.doi.org/10.1787/20792581; IEA (2014b), *Capturing the Multiple Benefits of Energy Efficiency: A Guide to Quantifying the Value Added*, International Energy Agency, Paris, http://dx.doi.org/10.1787/9789264220720-en; UNEP FI (2015), "Voluntary Energy Efficiency Investment Principles for G20 Participating Countries", www.unepfi.org/fileadmin/energyefficiency/EnergyEfficiencyInvestmentPrinciples.pdf (accessed 25 January 2016).

Box 4.2. Green investment banks mobilising green investment in cities

Many of the investments green investment banks (GIBs) mobilise are undertaken in urban areas, where 54% of the world's population lived in 2014 and where 66% is projected to live by 2050. For example, Australia's GIB, the Clean Energy Finance Corporation, is providing finance to help the city of Melbourne undertake an AUD 30 million programme of clean energy initiatives to help it reach its goal of zero net emissions by 2020. GIBs' energy efficiency activities focus particularly on buildings, which account for 19% of global GHG emissions.

Sources: UN DESA (2014), *World Urbanization Prospects: The 2014 Revision*, Highlights, United Nations Department of Economic and Social Affairs, Population Division, New York, http://esa.un.org/unpd/wup/Highlights/WUP2014-Highlights.pdf; IPCC (2014), "Summary for policymakers", in: Edenhofer, O. et al. (eds.), *Climate Change 2014: Mitigation of Climate Change, Contribution of Working Group III to the Fifth Assessment Report of the Intergovernmental Panel on Climate Change,* Cambridge University Press, Cambridge, United Kingdom and New York, New York, www.ipcc.ch/pdf/assessment-report/ar5/wg3/ipcc_wg3_ar5_summary-for-policymakers.pdf; CEFC (2015c), "Factsheet: CEFC and the city of Melbourne accelerate sustainability initiatives", October, **Error! Hyperlink reference not valid.** www.cleanenergyfinancecorp.com.au/media/107528/cefc-factsheet_cityofmelb_lr.pdf.

Financing investment in energy efficiency

The IEA valued global energy efficiency markets at between USD 310 billion and USD 360 billion in 2012, with high potential for growth (IEA, 2014a). Capital and savings from individuals, businesses and governments account for over half of this market. In order to achieve the full potential of energy efficiency, investment from other sources will be crucial.

Energy efficiency finance tools can be adapted to suit the investment needs and structures of various sectors and borrowers and can take the form of loans, bonds and equity investment. On-bill finance, performance contracting and leasing are also used as financing mechanisms.

While the private sector is a key energy efficiency investor, with commercial banks leading, the public sector plays an important role by catalysing additional private investment and improving energy efficiency in public buildings, state-owned industries and other public infrastructure (IEA, 2014a). Governments can reduce high transaction costs and risk by facilitating standardisation, creating loan warehouses or providing loan guarantees.

Public financial institutions (PFIs), discussed in Chapter 1 and Annex 1.A1, are investing significant amounts of capital in energy efficiency. An OECD working paper on "Public financial institutions and the low-carbon transition" provides case studies of five PFIs and highlights their role in mobilising investment in renewable energy and energy efficiency (Cochran et al., 2014). These institutions are generally much larger than GIBs, and as such, their investments in energy efficiency are also at a greater scale. For example, the German development bank KfW made an estimated EUR 6.5 billion in lending commitments for residential energy efficiency in 2011 (Cochran et al., 2014).[3]

Bilateral and multilateral development banks are also active in supporting energy efficiency investment through a range of interventions including making direct investments and providing risk mitigants and transaction enablers. For example, the Inter-American Development Bank (IDB) provided a USD 50 million line of credit to energy service companies which will originate and pool energy efficiency loans to Mexican small and medium-sized enterprises. The IDB will also provide up to USD 25 million in partial credit guarantees for the subsequent securitisation of the loan pool (IDB, 2014).

International climate finance funds such as the Climate Investment Funds (CIFs) and the Green Climate Fund (GCF) also target energy efficiency. For example, the European Bank for Reconstruction and Development (EBRD) and the Clean Technology Fund (CTF) launched a USD 350 million fund in 2014 to support residential energy efficiency programmes in Turkey through on-lending to local banks (Rosca, 2014).

Barriers to scaling up energy efficiency investment

Barriers to energy efficiency investment are generally well understood and often specific to the particular type of energy efficiency investment. High up-front costs are one type of barrier which is also common to renewable energy projects. Other barriers such as the principal-agent problem, where parties such as a landlord and tenant have different objectives and unequal access to information, are specific to energy efficiency investment (IEA, 2014a; 2014b). There are also a number of barriers that apply specifically to the finance element of energy efficiency investment (Box 4.3).

Box 4.3. Barriers to increasing the supply of energy efficiency finance

There are many barriers to energy efficiency investment. These barriers include the following:

- Small project size: Projects are often diffuse and too small to be attractive to lenders. As a result, project development and implementation costs are higher.
- Transaction costs: Companies may not apply for grant or loan programmes because filling out forms or reporting on energy savings is burdensome. Companies may also lack the technical expertise to implement energy efficiency projects. This is a significant factor affecting access to finance for businesses, particularly for small and medium-sized enterprises.
- Intangibility: Financial institutions may not consider energy savings (i.e. avoided energy costs) to be a potential source of cash flow that could be used for debt repayments. This is particularly problematic in industry, where a significant amount of savings can be achieved by altering processes rather than investing in new assets.
- Lack of harmonised monitoring and verification (M&V) protocols: Independent assessment of projects using M&V protocols is needed to win the trust of financiers, as energy savings typically change over time depending on production volumes, process changes and equipment degradation.
- Lack of data and skills to assess transactions and risk: A lack of transparent data and research makes it difficult to compare performance and attract investors. Performance data for energy efficiency projects are not collected systematically.
- Lack of financial instruments and funds with attributes that are attractive to institutional investors: Few available financial instruments and funds have the investment grade ratings, low transaction costs and liquidity that would be attractive to institutional investors.
- Policies and regulations that favour investment in unabated fossil-fuel intensive activities: Inconsistent policy signals, such as continued support for fossil fuel use, low or no carbon prices, and unpredictable changes to energy efficiency policies can limit the attractiveness of energy efficiency investments.
- Financial regulations with unintended consequences: International financial regulations to increase banks' level of capital and reduce their exposure to long-term debt may discourage long-term investments in areas such as energy efficiency.

Source: IEA (2014a), *Energy Efficiency Market Report 2014*, International Energy Agency, Paris, http://dx.doi.org/10.1787/9789264218260-en.

Role of green investment banks in energy efficiency investment

GIBs are joining other actors in efforts to promote private investment in energy efficiency projects. To capitalise on the opportunities energy efficiency presents to reduce energy consumption, lower GHG emissions and generate returns for private investors, several GIBs have launched energy efficiency programmes and invested in energy efficiency projects. Many GIBs also have a strategic mandate to promote job creation and economic growth. Energy efficiency investment can satisfy these mandates as projects typically require on-the-ground contractor labour and can spur business development. Table 4.1 highlights the energy efficiency investment offerings found at operational GIBs to date.

Table 4.1. **Green investment bank energy efficiency offerings**

Entity	Energy efficiency financing (target sector)
Clean Energy Finance Corporation (CEFC) (Australia)	– On-bill financing (commercial) – Efficiency fund (manufacturing) – Corporate financing (commercial) – Debt fund (local government) – Pass-through concessional loan (commercial equipment and vehicles)
Connecticut Green Bank (Connecticut, United States)	– Loan loss reserve (residential) – Credit enhancement (commercial, multi-family housing) – Property-assessed clean energy (PACE) origination and warehousing (commercial)
Malaysia Green Technology Corporation (GreenTech Malaysia) (Malaysia)	– Loan guarantees (commercial, municipal, universities, hospitals, schools)
NY Green Bank (New York, United States)	– Efficiency fund (commercial) – Equipment leasing (commercial) – Warehousing and credit enhancement (residential)
Technology Fund (Switzerland)	– Loan guarantees (innovative technologies)
UK Green Investment Bank (United Kingdom)	– Efficiency fund (non-residential including hospitals) – Corporate financing (municipal lighting)

A challenge facing GIBs in their effort to scale up energy efficiency investment is a lack of demand, as manifested by low uptake reported by some lenders that offer traditional energy efficiency products. GIBs report that when they ask banks to consider new or increased efficiency lending, banks sometimes assert that increased financing is unnecessary due to low demand. Commercial banks may deny the existence of a financing gap (and therefore the need for GIB interventions) based on their view that financing is available but unused. However, other factors may impact demand for financing.

The lack of demand for energy efficiency finance can be a result of insufficient or ineffective marketing efforts or a lack of co-ordination with contractor networks. With respect to marketing, successful energy efficiency financing programmes tend to be simple, are tailored to a target market and place minimal burden on the customer. Efficiency upgrades are rarely something a customer actively seeks, so an efficiency financing product is typically "sold" rather than "bought". With respect to contractors, successful financing products tend to be integrated with the efficiency service itself, easy to understand and repayable through minimal additional effort (e.g. through on-bill payments or tax payments). To increase demand, banks can inform contractors that they offer this financing product and encourage them to inform customers (DeVries, 2015).

Unattractive financing terms can also reduce demand for energy efficiency investments. Energy efficiency financing offered at high rates or with short tenors may impede cost-effective projects. In contrast, offering loan tenors and payment schedules and amounts to align with project savings can allow borrowers to save money on a monthly basis. For example, the UK Green Investment Bank has developed a tailored lending programme for local municipalities that is initially focused on street lighting upgrades and that matches payment schedules to when project savings are generated.

Successful GIB energy efficiency activities to date have highlighted the importance of designing "whole market" solutions and financing "whole building" efforts. "Whole market" solutions call for differentiated marketing and finance approaches for different segments of the local efficiency market, such as new buildings, buildings to be renovated, government-owned buildings, residential homes, commercial properties, etc. A "whole building" approach considers synergies involving efficiency upgrades and other renovations in buildings. For example, the Connecticut Green Bank has found that commercial buildings scheduled for renovation for some other purpose are the ideal target for energy efficiency lending. In addition, a majority of buildings suitable for energy efficiency investing are also good candidates for rooftop solar or combined heat and power (CHP) facilities.

GIBs understand that without origination of projects, very little market activity will occur. To provide advice and one-stop shopping for residential and commercial building owners, a GIB operating at the retail level may design an integrated financing and efficiency offering in partnership with local lenders and origination firms. Alternatively, a GIB may provide its own origination services or may on-lend to firms that in turn extend loans to building owners.

Some GIBs also facilitate market development by using aggregation techniques to build portfolios of similar loans large enough to attract private investors. For example, the Connecticut Green Bank issues energy efficiency loans to individual projects for commercial buildings and sells the cumulative portfolio to private investors in order to recapitalise the pool (US Department of Energy, 2014). Other GIBs like NY Green Bank and Australia's Clean Energy Finance Corporation (CEFC) have taken a wholesale approach, offering warehouses and lines of capital that project developers and lenders can draw upon to directly finance projects. Both approaches help to overcome investment barriers associated with small and disparate projects.

To facilitate pooling or bundling of loans, GIBs can increase standardisation by creating consistent loan documentation and technical assessment processes. The wide range of documentation, processes and project types can make it costly for investors to underwrite energy efficiency loans and also inhibits the creation of secondary markets for energy efficiency. Selling a portfolio of loans, either through private placement or public securitisation, requires a certain level of consistency across the loan within the portfolio. GIBs can promote standardisation in their deals to create greater uniformity across the market (Lowder and Mendelsohn, 2013).

Risk mitigation techniques are often used for energy efficiency projects that have very low project risk but still require additional support to make private lenders comfortable to participate in the projects. Risk mitigation may be provided through subordinated debt investments, loan loss reserves or loan guarantees. GreenTech Malaysia provides loan guarantees for energy efficiency investments, with varying fee schemes (GTFS, 2014). The Connecticut Green Bank's Smart-E Loan Program offers a standard loan loss reserve to local banks that make residential energy efficiency loans. In exchange for use of the reserve, lenders must provide loans below a maximum rate and longer than a minimum term.

GIBs employ these strategies through a range of different financing structures. A set of increasingly common techniques is used to overcome barriers and facilitate energy efficiency investment, while fulfilling the GIBs' mission of expanding efficiency markets by leveraging private investment. The following sections and Table 4.2 highlight the types of GIB energy efficiency investments and risk-mitigating and transaction-enabling offerings.

Table 4.2. **Types of green investment bank energy efficiency investments by entity**

Entity	On-bill financing	PACE	Credit enhancements	Efficiency funds	Direct lending	Leasing	Warehousing
Clean Energy Finance Corporation (CEFC) Australia	X	X		X	X	X	X
Connecticut Green Bank Connecticut, United States	In development	X	X		X		X
Malaysia Green Technology Corporation (GreenTech Malaysia) Malaysia			X				
NY Green Bank New York, United States			X	X		X	X
Technology Fund Switzerland			X				
UK Green Investment Bank United Kingdom				X	X		

Energy efficiency instruments and funds

In addition to making private investment less risky and costly, GIBs also directly invest public capital in energy efficiency projects. GIBs use a range of investment financing instruments, including dedicated efficiency funds, direct loans, equipment leases and warehousing for smaller efficiency loans. By directly investing, GIBs create opportunities for private sector co-investment, thereby mobilising private investment into underserved markets.

Energy efficiency funds

GIBs have created numerous project-based funds to provide loans to or otherwise support energy efficiency projects (UK Green Investment Bank, 2014a; CEFC, 2015a). GIBs may act directly as the marketer and underwriter for each loan, or partner with a private actor who is responsible for finding projects and disbursing loans. These funds demonstrate to the market that energy efficiency investments can be profitable and provide sufficient size to attract larger investors, such as investment banks or institutional investors.

In February, 2014 the UK Green Investment Bank formed a GBP 50 million energy efficiency partnership with Société Générale Equipment Finance, with each party committing GBP 25 million. The partnership will provide loans for CHP plants, boilers, building retrofits, lighting or energy reduction technologies for production processes. Loans will be structured so that repayments are less than the value of energy savings, meaning borrowers can save money on day one of the loan (UK Green Investment Bank, 2014a). Similarly, Australia's CEFC formed an energy efficiency fund with Commonwealth Bank, with each party investing AUD 50 million. The fund will make individual loans in the range of AUD 500 000-5 million, aimed at reducing energy costs (CEFC, 2015a).

Corporate finance for efficiency products

GIBs may provide direct corporate financing to companies to undertake energy efficiency projects and upgrades on their own buildings. Though corporations may recognise that investing their own resources in energy efficiency can create a positive return, other capital projects are typically given higher priority. Direct corporate finance addresses this challenge by enabling companies to implement projects without using their own capital. Australia's CEFC has directly financed a range of projects including: improving energy efficiency in lighting, heating and insulation at a local aquatic centre; cutting energy consumption through improved ventilation and LED lighting in a Brisbane office block; and helping a local council building reduce energy costs through improved air conditioning and energy-efficient lighting (CEFC, n.d.). The corporate loan facility funded by the UK Green Investment Bank to promote street lighting energy efficiency is discussed in Box 4.4.

Box 4.4. UK Green Investment Bank's municipal street-lighting loan

There are over 7 million street lights in the United Kingdom which generate over GBP 300 million in electricity costs. The electricity needed to power street lights produces 1.3 million tonnes of CO_2 annually, equivalent to the emissions of 330 000 cars on the road or 674 000 households. Despite the financial and environmental case for improved energy efficiency, fewer than 1 million street lamps are energy efficient.

To encourage municipalities to make the switch to low-energy lighting, the UK Green Investment Bank created an innovative "Green Loan" product in 2014 for municipalities which is specifically tailored to help cities upgrade their street lighting to more energy-efficient light emitting diodes (LEDs). The efficient lighting technology produces energy savings that exceed the cost of the loan payment, allowing borrowers to be cash-flow positive throughout the period of the loan. The product's fixed rates and terms designed to match the payback period allow cities and towns to enjoy net savings on their street lighting from day one of the project and municipalities save 80% of their lighting costs. By using this product, participating municipalities reduce their operating budgets and take advantage of investment opportunities that otherwise would be left untapped because of competing investment needs deemed to be of higher priority.

Source: UK Green Investment Bank (2014b), "Low energy streetlighting: Making the switch", Market Report, UK Green Investment Bank, February, www.greeninvestmentbank.com/media/5243/gib-market-report-low-energy-streetlighting-feb-2014-final.pdf.

In another example, Australia's CEFC has financed National Australia Bank via a corporate bond purchase in exchange for offering a concessional loan product for financing equipment and vehicles that meet CEFC standards of efficiency. The "Energy Efficient Bonus" is offered to the end user as a 70 basis point (0.7%) discount from the prevailing equipment finance rate. This provides equipment sales persons with a talking point about energy efficiency and entices the purchaser to compare the costs of a more efficient product with the costs of less efficient products that do not qualify for the bonus (CEFC, 2015b).

Equipment leasing

Through an equipment lease, the lessor maintains ownership and the lessee makes regular payments. The lessee gets the benefits of using the new equipment – in this case the reduced energy cost – without having to use internal resources to pay for the equipment upfront. This financing method allows borrowers looking for new energy-efficient equipment to replace one operating expense (energy) with another (lease payments) without making a capital expenditure. The corporate balance sheet is therefore unaffected by the lease and there may also be tax benefits from lease payment deductions (NRDC, 2011).

In 2014, NY Green Bank announced the creation of an energy equipment financing fund. In partnership with Bank of America Merrill Lynch, NY Green Bank announced an agreement in principle to co-invest long-term capital aimed at expanding commercial market offerings for equipment leasing and enabling deeper energy retrofits (NY Green Bank, 2014). The fund will finance public and private sector renewable energy projects, including renewable energy, energy efficiency and CHP.

Risk mitigants and transaction enablers for energy efficiency investment

GIBs are engaged in a range of activities to reduce the risk of energy efficiency investments or to help lower the high transaction costs often associated with energy efficiency projects. Investment in energy efficiency projects is often unsecured, with the lender unable to claim ownership of a physical asset in the event of default. Due to the lack of collateral and associated risks, interest rates may be high, reducing the attractiveness of energy efficiency loans. To address this challenge, several GIBs use new mechanisms such as on-bill finance and property-assessed clean energy (PACE) financing, which allow energy efficiency loans to be paid back through utility bills or property taxes, reducing repayment risk. GIBs can also offer traditional risk mitigants such as loan guarantees or first-loss provisions.

Guarantees

Malaysia's Green Technology Financing Scheme (GTFS) offers loan guarantees to energy efficiency projects. Properly certified green projects may seek loans from participating commercial banks, which in turn receive a 60% loan repayment guarantee from the GTFS. A broad range of energy efficiency technologies and solutions are eligible for the guarantee (GTFS, 2012). In addition to guaranteeing loan repayment, GIBs could (in principal) guarantee the energy savings achieved through an efficiency project to increase consumer confidence and spur investment.

On-bill finance

In the most basic on-bill finance structure, a lender issues a loan to a borrower for an energy efficiency project. Instead of having the lender send a loan repayment bill to the borrower, the cost of repayment is listed directly on the monthly energy utility bill the borrower already receives. The utility collects payment from the borrower and remits the payment back to the lender. This technique is attractive for the borrower because bills are consolidated and the borrower can see on a single bill the reduced energy expenditure and the corresponding cost of the loan repayment. It is attractive for lenders due to the low historical default rate of utility bills compared to unsecured consumer financing (State and Local Energy Efficiency Action Network, 2014). The additional repayment security can lower the interest rate on the loan, as in the case of Australia's CEFC's on-bill finance programme described in Chapter 3 (Origin, n.d.).

On-bill finance can be enhanced to provide even greater lender security by creating a tariff-based on-bill programme. Using this approach, the loan is tied to the utility meter of the building, not the individual borrower, so if a property is sold the loan stays with the building instead of the individual owner. This provides greater transparency for building sales and eliminates potential borrower concerns regarding cost recovery if a property is sold.

GIBs can support and facilitate on-bill finance by acting as the primary lender of an on-bill programme, aggregating a portfolio of loans that can then be sold to a private investor. GIBs could also co-invest with a private lender or provide a credit enhancement like a loan loss reserve to support loan origination provided directly by a private investor. GIBs could also play an administrative role, helping to establish a new on-bill programme by co-ordinating lenders, policy makers and regulators.

Property-assessed clean energy

Property-assessed clean energy, or "PACE", is a form of renewable energy financing through which a borrower repays a loan through property taxes attached to the building that is being upgraded. When a PACE loan is issued, a new property tax lien is placed on the building that benefits from the energy efficiency improvement. By creating a lien, the loan repayment is treated like a new tax obligation on the borrower, with the building itself used as collateral. If a building owner does not pay their property taxes, the government can foreclose on the building and sell it in order to recover the unpaid tax obligation. Under PACE, the efficiency loan is treated the same way, with a penalty of foreclosure in the event the borrower does not repay the efficiency loan. This is a powerful tool to be applied to energy efficiency financing, which is typically why PACE statutes require PACE projects to be cash-flow positive from the start of the loan term (NREL, 2010).

Much like on-bill financing, the lien makes repayment effortless for borrowers and creates increased security for lenders. Rather than treating an energy efficiency loan as unsecured consumer or corporate debt, banks can treat the PACE structure as a far more secure repayment which enables lower rates and longer term lending and attracts new investors. PACE financing is most commonly used in the United States.

PACE financing programmes typically require enabling legislation and can be complex to implement. Many local jurisdictions have a long history of using tax liens and "special improvement districts" to facilitate investment in public infrastructure with repayment through property taxes. If, for example, a city decides to improve the local sewage system, it may issue a bond to pay for the improvement. To recover project costs, the city could identify the "special improvement district" and place a new property tax lien on all properties that benefit from the system. To enable PACE programmes, legislation must be passed that allows local tax-collecting jurisdictions to treat renewable energy investment like other infrastructure which can be repaid through tax liens. A given jurisdiction must then opt to allow the placement of these liens within its property tax base. PACE liens, unlike other infrastructure liens, are entirely voluntary and are only placed on buildings that receive a renewable energy loan.

PACE programmes can be difficult to structure as they require legal authorisation and close co-ordination between lenders, local governments, programme administrators and contractors. In many states in the United States this complexity has hindered market growth. While many US states have passed PACE-enabling legislation, the tool is only used at scale in California for residential PACE and in Connecticut for commercial PACE. Growth is slowed not by lack of consumer demand but by inefficient legal and programme structures that place the burden on each local jurisdiction to create their own PACE programme with independent financing sources. The Connecticut Green Bank, however, has found notable success by centrally administering and financing a state-wide commercial energy efficiency programme (see Box 4.5 for details). Its "C-PACE" programme co-ordinates all commercial PACE activity in the state, originating loans with public capital and then selling the portfolio of loans to private investors. The first portfolio sale of USD 30 million in early 2014 represented the first securitisation of its kind. Other US states are exploring the use of GIB financing to create similar commercial PACE programmes.

Reducing transaction costs through warehousing

Warehousing can be an element in many forms of energy efficiency financing, paired with tools such as PACE, on-bill financing or equipment leases. For example, the Connecticut Green Bank warehouses its PACE loans, which are then sold as a whole portfolio to private investors (described in Box 4.5). By using the warehouse structure, the Connecticut Green Bank is able to create consistency and address many small projects, and then aggregate them to a scale that is attractive for private investors.

Box 4.5. **The Connecticut Green Bank C-PACE programme**

The Connecticut Green Bank has implemented one of the most successful commercial building energy efficiency programmes in the United States, using the property-assessed clean energy (PACE) structure. The programme was launched in early 2013 and in less than two years the Green Bank has financed nearly USD 54 million in energy upgrades for 89 buildings. This accounts for about one-third of the commercial PACE market in the United States. More recently, the Green Bank has established a programme to facilitate private platforms to provide PACE financing, with the Green Bank retaining its central administration role. Other US states such as Rhode Island are exploring the use of a green investment bank (GIB) to facilitate similar commercial PACE programmes.

Connecticut is one of 29 US states to pass PACE-enabling legislation, but it is the only one to have created a state-wide programme with centralised administration through a GIB. This structure was created to avoid the pitfalls of relying on individual jurisdictions to each create distinct programmes, guidelines and financing strategies. The Connecticut Green Bank provides a standardised approach for all commercial PACE deals in the state, allowing for greater scale. In addition, the Green Bank committed to educating municipalities, contractors, building owners and private banks about the programme. Many other states have found that private actors are slow to take on this public goods-generating role. Connecticut's legislation also tasked the Green Bank with certifying that each PACE deal in the state meets a certain level of quality, as measured by the savings-to-investment ratio of the project. Each PACE project must have a savings-to-investment ratio above 1.0, meaning all projects must be cash-flow positive.

As originally designed, the Green Bank intended to establish the PACE programme structures and then invite private lenders to originate loans, with the Green Bank co-ordinating with municipalities and approving deals. However, despite pre-approving multiple banks to participate in the programme, private lenders did not enter the market after the programme was launched as they were still hesitant to be the first investors in a new and unfamiliar structure. This led the Green Bank's Board of Directors to authorise the origination of PACE loans using its own balance sheet and the creation of an internal USD 40 million warehouse which could be used to originate loans through contractor networks and direct marketing. Loans are currently issued at 5-6% rates, with terms up to 20 years, intended to match the useful life of the energy conservation measures. The Green Bank performs financial underwriting for each deal, and partners with a technical administration firm that produces detailed technical assessments, savings projections and return calculations to facilitate deal closing and approval.

Box 4.5. **The Connecticut Green Bank C-PACE programme** *(continued)*

Figure 4.1. **Connecticut Green Bank's C-PACE enables secure efficiency investment at scale**

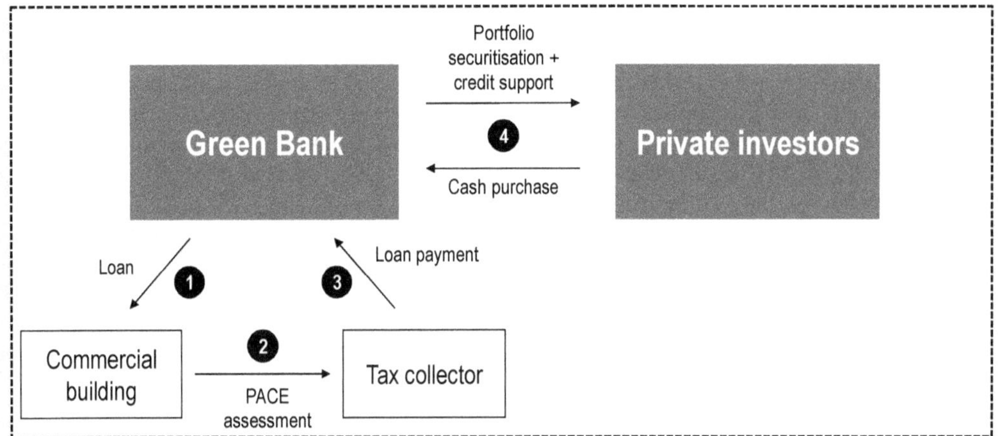

To leverage private capital and recapitalise the warehouse, in early 2014, the Green Bank invited bids from investors interested in purchasing the PACE loans. After achieving the necessary scale and creating project consistency, the offer attracted bids from numerous firms, including Clean Fund, a niche PACE investment firm and the eventual bid winner. Under the deal terms, Clean Fund bought USD 24 million of a USD 30 million bond issuance, with the Green Bank repurchasing USD 6 million worth of bonds in a subordinate tranche as a credit enhancement to Clean Fund. The deal marked the first commercial energy efficiency securitisation of its kind and serves as a model for how GIBs can overcome barriers to finance, bring markets to scale and attract private capital. The Green Bank is now seeking to build a larger external origination warehouse funded with both Green Bank and private capital, with a goal of selling the PACE loans through a public securitisation.

Sources: Lombardi, N. (2014), "In a 'watershed' deal, securitization comes to commercial efficiency", Greentech Media, 19 May, www.greentechmedia.com/articles/read/the-first-known-commercial-efficiency-securitization; Connecticut Green Bank (2015), "Innovating, educating and activating to accelerate clean energy: 2014 annual report", Connecticut Green Bank, Stamford, Connecticut, www.ctgreenbank.com/wp-content/uploads/2015/12/AnnualReport_FINAL_5.4.15-SinglePages.pdf; Coalition for Green Capital (2015), "Creating state financing tools to make clean energy markets grow quickly", presentation by Reed Hundt, May, www.coalitionforgreencapital.com/uploads/2/5/3/6/2536821/cgc_-_summary_presentation_may_2015.pdf; PACE Now (n.d.), "List of all PACE enabling statutes by state," PACE Now website, www.pacenow.org/resources/pace-enabling-legislation (accessed 25 August 2015).

Energy efficiency investment partners

Given the range of barriers preventing the scaling up of energy efficiency investment, GIBs engage with a range of partners to mobilise private investment in energy efficiency, including local banks, retail energy efficiency firms and local development authorities.

Box 4.6. **Warehouse for Energy Efficiency Loans (WHEEL)**

The Warehouse for Energy Efficiency Loans (WHEEL) is a cross-state energy efficiency financing platform launched in the United States to attract institutional investors by achieving scale through aggregation of projects and consistency through project standardisation. Based on a programme started in Pennsylvania, WHEEL provides a credit enhancement to a centralised, privately-funded, national warehouse, which, in exchange, provides capital to fund energy efficiency loans in that state. This structure allows each state to design its own deployment and retail lending strategy while taking advantage of low-cost institutional capital drawn from the national warehouse. The initial investors Citi and Renewable Funding have built a USD 100 million loan pool, which will be securitised and recapitalised once funds are fully deployed. The first WHEEL securitisation of USD 12.58 million backed by pools of residential energy efficiency loans took place in June 2015 with plans to execute additional transactions in the next several years. Pennsylvania and Kentucky were charter members of WHEEL, and in October 2014 WHEEL expanded into New York through a NY Green Bank investment. As per the requirements of WHEEL, NY Green Bank offered a credit enhancement to the central loan fund, allowing New York borrowers to access the warehouse.

Sources: PR Newswire (2014), "U.S. homeowners to benefit from groundbreaking home energy loan financing platform", 9 April, www.prnewswire.com/news-releases/us-homeowners-to-benefit-from-groundbreaking-home-energy-loan-financing-platform-254545821.html; NY Green Bank (2014), "NY Green Bank's initial transactions," NY Green Bank website, http://greenbank.ny.gov/initial-transactions; Citi (2015), "Citi and Renew Financial announce first-ever energy efficiency loan asset-backed security transaction", press release, 15 June, Citigroup Inc., www.citigroup.com/citi/news/2015/150615a.htm.

Local banks

Building partnerships with smaller local banks is important for GIBs, particularly for residential and small commercial energy efficiency projects which, by their nature, have relatively low upfront costs. Local banks in regions with higher demand for energy efficiency projects, based on local climate and energy infrastructure, can play an important role in originating energy efficiency loans for individuals. They are also well positioned to pair energy efficiency loans with other forms of home lending, like mortgages and home equity loans. Home purchase or remodelling are ideal decision points for homeowners to consider energy efficiency upgrades, so positioning local banks to offer efficiency-specific lending products in tandem with a mortgage could prove an effective strategy for increasing consumer demand (Energy Star, n.d.).

GIBs can take on the role of originating small, disparate loans for a range of types of energy efficiency projects, and they can also work with local banks to take on the role of underwriting and originating loans. GIBs can support local banks with technical assistance or training, and also through financial support like credit enhancements or co-lending to incentivise local lending activity.

A barrier to increased local lending in energy efficiency is that local banks are often unfamiliar with or averse to unsecured lending that is paid back through energy savings. Local banks primarily issue mortgages and other home-equity based lending, or business loans to expand or improve local businesses. For these types of lending, risk assessment is based primarily on an individual borrower's credit worthiness and the perceived likelihood that income sources will be great enough to pay back the loan. Energy efficiency projects, despite creating their own income stream through energy savings, are frequently assessed using the same risk considerations as other loans that do not produce income to repay the loan. Because local banks do not account for project savings in the underwriting process, they treat efficiency loans as they do other loans that have greater risk of default (Schopp, 2014).

GIBs can work with local lenders to educate them on the nature of efficiency payback and help them develop more appropriate underwriting criteria that account for project savings. This can increase lenders' comfort with this kind of loan. GIBs can also drive market entry by local lenders through credit enhancements. As described earlier, the Connecticut Green Bank's Smart-E Loan Program targets local banks by offering a loan loss reserve to support energy efficiency loans. By enrolling in the programme, local lending institutions are ensured that the reserve would cover a portion of late payments and defaults. In exchange, banks agree to comply with "not-to-exceed" rates and offer better terms to borrowers than they would without the reserve. The Connecticut Green Bank also informs contractor networks on which banks are participating in order to drive customer demand to the banks offering Smart-E loans. In Malaysia, the GTFS's loan guarantee structure necessitates the participation of other private lenders to originate loans.

Retail efficiency firms

In addition to traditional local lending institutions, there are a growing number of niche private investment firms that exclusively focus on providing retail energy efficiency loans paired with direct energy efficiency services. These "one-stop shop" firms are often funded by institutional investors and provide an integrated solution tailored to market segments too small to attract large ESCO firms. For example, Renovate America operates residential PACE financing programmes throughout California, with large loan portfolios securitised through public sales (Hales, 2015). GIBs can partner with or support these private firms by creating funds or providing a credit enhancement to enable the private firm to target customers with lower incomes or credit rating.

Local development authorities

Sub-national governments may have associated development authorities which have the power to issue bonds to support infrastructure projects. These authorities can directly access low-cost debt in public markets based on the backing of sub-national government credit. GIBs can work with development authorities to identify private investment partners, help structure deals, identify energy project opportunities and create sustained energy finance programmes. For example, the Port of Greater Cincinnati Development Authority in the state of Ohio issued bonds to finance the local PACE programme (Port of Greater Cincinnati Development Authority, 2015).

Notes

1. In 2011, energy savings from efficiency improvements in 11 IEA member countries was greater than the energy consumed by any single energy source across those countries (IEA, 2014a).

2. In the IEA's *World Energy Outlook 2012* (IEA, 2012), investments were classified as competitive if the payback period for the up-front energy efficiency investment is equal to or less than the amount of time an investor might be reasonably willing to wait to recover the cost, using the value of undiscounted fuel savings as a metric. The payback periods used were (in some cases) longer than current averages, but they were always shorter than the technical lifetime of individual assets.

3. The EUR 6.5 billion commitment led to EUR 18.4 billion in total investments in 282 000 households. The cost to the federal budget was EUR 934 million (for funding for subsidised lending, repayment bonuses, etc.). This represents a leverage effect of nearly EUR 20 of private investment for every EUR 1 of public capital (Cochran et al., 2014). However, methodologies for calculating leverage ratios may differ and caution should be used when comparing relative mobilisation rates across institutions.

References

CEFC (2015a), "Support for manufacturers, small business with energy efficient loans," Fact Sheet, Clean Energy Finance Corporation, Sydney, Australia, www.cleanenergyfinancecorp.com.au/media/76221/cefc-factsheet-energyefficientloans_lr.pdf.

CEFC (2015b), "CEFC and the city of Melbourne accelerate sustainability initiatives", Fact Sheet, October, Clean Energy Finance Corporation, Sydney, Australia, www.cleanenergyfinancecorp.com.au/media/107528/cefc-factsheet_cityofmelb_lr.pdf.

CEFC (2015c), "Factsheet: CEFC and the city of Melbourne accelerate sustainability initiatives", October, **Error! Hyperlink reference not valid.** www.cleanenergyfinancecorp.com.au/media/107528/cefc-factsheet_cityofmelb_lr.pdf.

CEFC (n.d.), "Investment case studies", Clean Energy Finance Corporation website, www.cleanenergyfinancecorp.com.au/investments/case-studies.aspx (accessed 1 September 2015).

Citi (2015), "Citi and Renew Financial announce first-ever energy efficiency loan asset-backed security transaction", press release, 15 June, Citigroup Inc., www.citigroup.com/citi/news/2015/150615a.htm.

Coalition for Green Capital (2015), "Creating state financing tools to make clean energy markets grow quickly", presentation by Reed Hundt, May, www.coalitionforgreencapital.com/uploads/2/5/3/6/2536821/cgc_-_summary_presentation_may_2015.pdf.

Cochran, I. et al. (2014), "Public financial institutions and the low-carbon transition: Five case studies on low-carbon infrastructure and project investment", *OECD Environment Working Papers*, No. 72, OECD Publishing, Paris, http://dx.doi.org/10.1787/5jxt3rhpgn9t-en.

Connecticut Green Bank (2015), "Innovating, educating and activating to accelerate clean energy: 2014 annual report", Connecticut Green Bank, Stamford, Connecticut, www.ctgreenbank.com/wp-content/uploads/2015/12/AnnualReport_FINAL_5.4.15-SinglePages.pdf.

DeVries, C. (2015), "Energy efficiency finance demand isn't the problem; design is", State & Local Energy Report, 12 February, http://stateenergyreport.com/2015/02/12/energy-efficiency-finance-demand-isnt-the-problem-design-is.

Energy Star (n.d.), "Energy efficient mortgages", Energy Star website, www.energystar.gov/index.cfm?c=mortgages.energy_efficient_mortgages.

GTFS (2014), "Green Technology Financing Scheme", presentation by Tan Ching Tiong at the UNFCCC Regional Workshop for the Asia-Pacific Region of NAMAs, 22-25 April, Vientiane, Lao People's Democratic Republic, https://unfccc.int/files/focus/mitigation/application/pdf/malaysia_presentation.pdf.

GTFS (2012), "Criteria for building and township sector", Green Technology Financing Scheme, GreenTech Malaysia, www.gtfs.my/page/criteria-building-and-township-sector.

Hales, R.L. (2015), "Renovate America's 5th securitization of PACE bonds", Clean Technica, 3 December, http://cleantechnica.com/2015/12/03/renovate-americas-5th-securitization-pace-bonds.

IDB (2014), "Capital markets solution for energy efficiency financing", *Environmental and Social Management Report*, ME-I1150, Inter-American Development Bank, www.iadb.org/en/projects/project-description-title,1303.html?id=ME-L1150.

IEA (2014a), *Energy Efficiency Market Report 2014*, International Energy Agency, Paris, http://dx.doi.org/10.1787/9789264218260-en.

IEA (2014b), *Capturing the Multiple Benefits of Energy Efficiency: A Guide to Quantifying the Value Added*, International Energy Agency, Paris, http://dx.doi.org/10.1787/9789264220720-en.

IEA (2012), *World Energy Outlook 2012*, OECD Publishing, Paris, http://dx.doi.org/10.1787/weo-2012-en.

IPCC (2014), "Summary for policymakers", in: Edenhofer, O. et al. (eds.), *Climate Change 2014: Mitigation of Climate Change, Contribution of Working Group III to the Fifth Assessment Report of the Intergovernmental Panel on Climate Change*, Cambridge University Press, Cambridge, United Kingdom and New York, New York, www.ipcc.ch/pdf/assessment-report/ar5/wg3/ipcc_wg3_ar5_summary-for-policymakers.pdf.

Lombardi, N. (2014), "In a 'watershed' deal, securitization comes to commercial efficiency," Greentech Media, 19 May, www.greentechmedia.com/articles/read/the-first-known-commercial-efficiency-securitization.

Lowder, T. and M. Mendelsohn (2013), "The potential of securitization in solar PV finance", Technical Report NREL/TP-6A20-60230, National Renewable Energy Laboratory, US Department of Energy, Office of Energy Efficiency & Renewable Energy, December, www.nrel.gov/docs/fy14osti/60230.pdf.

NRDC (2011), "Energy Efficiency Lease Guidelines", Issue Paper, November, Center for Market Innovation, Natural Resources Defense Council, www.nrdc.org/greenbusiness/cmi/files/CMI-FS-Energy.pdf.

NREL (2010), "Property assessed clean energy (PACE) financing of renewables and efficiency", Fact Sheet Series on Financing Renewable Energy Projects, NREL/BR-6A2-47097, NREL Energy Analysis, National Renewable Energy Laboratory, Golden, Colorado, July, www.nrel.gov/docs/fy10osti/47097.pdf.

NY Green Bank (2014), "NY Green Bank's initial transactions," NY Green Bank website, http://greenbank.ny.gov/initial-transactions.

Origin (n.d.), "Leading Australia in energy efficiency solutions", Factsheet, Origin and Low Carbon Australia, www.cleanenergyfinancecorp.com.au/media/40801/origin_factsheet.pdf.

PACE Now (n.d.), "List of all PACE enabling statutes by state," PACE Now website, www.pacenow.org/resources/pace-enabling-legislation (accessed 25 August 2015).

Port of Greater Cincinnati Development Authority (2015), "Port Authority expands regional financing options", www.cincinnatiport.org/port-authority-expands-regional-financing-options.

PR Newswire (2014), "U.S. homeowners to benefit from groundbreaking home energy loan financing platform", 9 April, www.prnewswire.com/news-releases/us-homeowners-to-benefit-from-groundbreaking-home-energy-loan-financing-platform-254545821.html.

Rosca, O. (2014), "EBRD rolls out US$ 350 million home energy efficiency programme in Turkey", European Bank for Reconstruction and Development, 19 November, www.ebrd.com/news/2014/ebrd-rolls-out-us-350-million-home-energy-efficiency-programme-in-turkey.html.

Ryan, L. and N. Campbell (2012), "Spreading the net: The multiple benefits of energy efficiency improvements", *IEA Energy Papers*, No. 2012/08, OECD Publishing, Paris, http://dx.doi.org/10.1787/20792581.

Schopp, A. (2014), "Incentives for commercial banks to finance energy efficiency", German Institute for Economic Research (DIW Berlin), www.diw.de/documents/vortragsdokumente/220/diw_01.c.471701.de/v_2014_schopp_incentives_eea.pdf.

State and Local Energy Efficiency Action Network (2014), "Financing energy improvements on utility bills: Market updates and program design considerations for policymakers and administrators", DOE/EE-1100, State and Local Energy Efficiency Action Network, May, www4.eere.energy.gov/seeaction/system/files/documents/onbill_financing.pdf.

UK Green Investment Bank (2014a), "Green Investment Bank Forms £50m alliance with SGEF to finance energy efficiency measures", United Kingdom Green Investment Bank, 25 February, www.greeninvestmentbank.com/news-and-insight/2014/green-investment-bank-forms-50m-alliance-with-sgef-to-finance-energy-efficiency-measures.

UK Green Investment Bank (2014b), "Low energy streetlighting: Making the switch", Market Report, UK Green Investment Bank, February, www.greeninvestmentbank.com/media/5243/gib-market-report-low-energy-streetlighting-feb-2014-final.pdf.

UN DESA (2014), *World Urbanization Prospects: The 2014 Revision*, Highlights, United Nations Department of Economic and Social Affairs, Population Division, New York, http://esa.un.org/unpd/wup/Highlights/WUP2014-Highlights.pdf.

UNEP FI (2015) "Voluntary Energy Efficiency Investment Principles for G20 Participating Countries", www.unepfi.org/fileadmin/energyefficiency/EnergyEfficiencyInvestmentPrinciples.pdf (accessed 25 January 2016).

US Department of Energy (2014), "Getting to scale: Energy investment partnerships & early secondary market transactions", US Department of Energy Better Buildings Summit, http://betterbuildingssolutioncenter.energy.gov/sites/default/files/getting_to_scale_energy_investment_partnerships.pdf.

Chapter 5.

Setting up and capitalising a green investment bank

This chapter provides "nuts and bolts" information regarding the process of setting up a green investment bank. It introduces the political processes that may be pursued to establish a green investment bank and discusses sources of capitalisation and continued funding. The importance of appropriate leadership, staffing and oversight is discussed as well as the variety of reporting and evaluation metrics used by green investment banks.

Key takeaways

- Green investment banks can be established using a variety of government processes including executive action, regulation and legislation, depending on the country context and the intended legal and administrative structure of the organisation.
- Some green investment banks are independent of government or quasi-independent, while others may exist as independent units or funds operating within government.
- Numerous financial sources, such as government appropriations, utility bill charges or carbon tax revenues, can be used to capitalise a green investment bank. Reallocated resources from existing programmes can also be a source of funds.
- Given the demanding mandates green investment banks face, leadership and staff need deal-making expertise and strong communication skills to engage effectively with potential investors, auditors, other government agencies and the public.
- Accountability, independence, public transparency and oversight are essential for green investment banks. They are typically overseen by independent governing boards and report their performance against a variety of metrics, such as private capital mobilisation, job creation and greenhouse gas emission reductions.

Creating a green investment bank

Building momentum and a case for a green investment bank

To build a case for creating a green investment bank (GIB), local and national leaders typically call for the establishment of a GIB and for a feasibility study to be undertaken. Annex 5.A1 provides additional details regarding the processes under which GIBs were designed and created in Australia, New York (United States) and the United Kingdom. Common activities used by prospective GIBs in the early stage of the design and formation process include the following:

- Engaging external consultants and dedicated resources: External consultants have frequently been engaged to assist in the process of sizing the market, identifying barriers and conducting quantitative analysis of the impact of prospective GIB interventions. For example, the UK Department for Business, Innovation & Skills commissioned Vivid Economics, McKinsey and Deloitte to prepare a report on the rationale and costs and benefits of the proposed GIB (Vivid Economics and McKinsey & Co, 2011). The report also explored the GIB's value for money and analysed the economic equity and efficiency of a GIB versus alternative policy options in key sectors, such as increasing feed-in tariffs or landfill taxes. NY Green Bank hired Booz & Company for a detailed qualitative and quantitative analysis of the market, prospective interventions and impact (Booz & Company, 2013). In Australia, an Expert Review Panel was tasked with preparing a detailed study of the prospective institution (Australian Government, 2012).
- Studying local and international experiences: GIBs carefully study peer organisations and fellow GIBs. For example, NY Green Bank studied existing GIBs and GIB-like institutions, including the UK Green Investment Bank, Germany's KfW and numerous US state-based programmes. In Australia, the

Expert Review team tasked with evaluating the GIB drew on international experiences including the UK Green Investment Bank, KfW, the Brazilian Economic and Social Development Bank (BNDES), the China Development Bank and loan programmes offered by the US Department of Energy.

- Using public comment periods: Public comment periods are common during the GIB formation process and submitted comments are often available online for consultation. For example, in Australia, the Expert Panel consulted widely with industry and stakeholders and received 170 submissions and 200 emails regarding the Clean Energy Finance Corporation's (CEFC's) potential scope, how it could work with other government and market organisations, and how to identify and overcome key market gaps for low-emissions technologies (CEFC, 2011).

Legal pathway to green investment bank creation

In each case, GIBs have been created through a government process granting the organisation certain legal authorisations and access to capital sources. In most cases, GIBs have been created through a legislative process, with bills passed through representative bodies that defined the scope, financing tools and capital sources for the banks. Some GIBs, like NY Green Bank, were created through a combination of administrative action by the executive branch of government and a regulatory process (New York Public Service Commission, 2013). The exact path for the establishment of a GIB will depend on the legal structure of the new institution, whether the bank will be part of or take over an existing institution, the desired financing capabilities and the necessary method for tapping capital sources. If the government identifies an existing public entity that already has the authorisation to perform GIB financing activities and a viable capital source, legislation or regulatory action may not be needed. However, in most cases GIBs represent a new approach to government financing or draw upon new capital sources, and consequently require legal or regulatory action to be created.

Opposition to establishing a green investment bank

The creation of a GIB may be subject to opposition from both political actors and citizens. Opponents of creating a GIB may include existing government and non-profit entities that offer programmes to support renewable energy adoption. Typically these entities offer grants and subsidies using public funding and may view GIBs as duplicative or as competitors. Opposition can be particularly strong if a GIB aims to repurpose grant funding for its initial capital as groups may support the idea of a green bank but oppose the diversion of funds from existing programmes (ECRI, 2014; Proft, 2015).

To reduce opposition to the creation of GIBs, GIB proponents have actively sought to engage existing renewable energy entities and their supporters in the process of launching a GIB. For instance, NY Green Bank was capitalised with repurposed ratepayer funds that were previously directed to state grant programmes co-ordinated through its parent organisation, NYSERDA. To gain buy-in from the community of supporters for those programmes, NY Green Bank sought public comments during the regulatory process to create the institution (New York State, 2013). Those comments were incorporated into the order creating the bank, which ensured that the bank would aim to utilise public capital at least as effectively as the existing programmes. As noted above, Australia's CEFC also consulted widely with industry and stakeholders and received 170 comment submissions (CEFC, 2011).

Table 5.1. **Summary of mechanisms for green investment bank creation**

Entity	Creation mechanism	Legal structure
California CLEEN Center (California, United States)	Board of Director decision	Centre within existing public infrastructure bank
Clean Energy Finance Corporation (CEFC) (Australia)	Legislation	Independent entity – a body corporate and Commonwealth authority
Connecticut Green Bank (Connecticut, United States)	Legislation	Quasi-public independent entity created from an existing entity
Green Energy Market Securitization (GEMS) (Hawaii Green Infrastructure Authority) (Hawaii, United States)	Legislation (establishment) and regulatory (funding)	Public entity, housed within a state economic development entity
Green Fund (Japan)	Executive action	Entity within national environmental ministry
Malaysian Green Technology Corporation (GreenTech Malaysia) (Malaysia)	Executive action	Non-profit entity operated by national energy ministry
Masdar (United Arab Emirates)	Executive action	Subsidiary of national economic development company
Montgomery County Green Bank (Maryland, United States)	Legislation	*To be determined*
New Jersey Energy Resilience Bank (ERB) (New Jersey, United States)	Executive action	Public entity, created and staffed jointly by state energy and commerce offices
NY Green Bank (New York, United States)	Executive action (establishment) and regulatory (funding)	Public entity, subsidiary of state government energy office
Rhode Island Infrastructure Bank (RIIB) (Rhode Island, United States)	Legislation	Quasi-public entity formed by rebranding and expanding existing water finance agency
Technology Fund (Switzerland)	Legislation	Political instrument of national government, public/private steering committee
UK Green Investment Bank (United Kingdom)	Executive action (establishment as a public limited company (plc) and legislation (providing parliamentary control and creating bespoke funding power)	Independent entity wholly owned by government

Private lenders and banks may also initially oppose the creation of a GIB out of concern that the public financing will displace or "crowd out" private lending activity. Prospective GIBs may seek to address these concerns during the consultation process, including by noting that they are mandated to demonstrate that their involvement in a transaction mobilised investment that would not have otherwise occurred. They may also point to cases in which existing GIBs have stepped back from their lending activity when private lenders moved into a new market sector.

Using the term "green bank" or "green investment bank" itself may create opposition from stakeholders who are opposed to government spending to address climate change. In some jurisdictions, even the term "bank" may draw opposition from those opposed to public finance. In the US state of Vermont, for instance, many years of contentious debate over an unrelated "state bank" has led GIB advocates to avoid using the term "green bank" in development efforts (personal communication with Jeffrey Schub, Coalition for Green Capital, 17 January 2016).

Political opposition to existing GIBs can also emerge. Australia's CEFC provides an example of how political support for the existence or focus of a GIB can shift. Australia's former Prime Minister Tony Abbott sought to abolish the CEFC, introducing legislation to that effect that was twice defeated in the parliament (CEFC, 2015a; Liberal Party of Australia, 2013). In December 2015, after the change in Prime Minister, the CEFC received a new investment mandate indicating that the "CEFC must include a focus on supporting emerging and innovative renewable technologies and energy efficiency, such as large-scale solar, storage associated with large and small-scale solar, offshore wind technologies, and energy efficiency technologies for cities and the built environment" (CEFC, 2015a). The CEFC highlighted that the mandate is not retrospective and will not impact existing investments, and that to manage risks it would need to balance investments with higher inherent risk (e.g. new and emerging technologies) alongside those with lower inherent risk (e.g. investments in more mature technologies) (CEFC, 2015a). This experience highlights the tension between GIBs' interest in remaining independent and able to implement long-term strategies, and governments' interest in deciding on how best to use public funds in light of political priorities.

Administrative set-up and positioning

The administrative structure of a GIB is determined when the institution is created in law. A GIB's governance structures, oversight and internal processes depend on whether it is part of government, a private non-profit or a quasi-public institution. A GIB that is separated, at least partially, from government may be better suited to maintain its mission through changing political landscapes. However, it can be difficult to pass a new law that allocates significant public funding to a new and independent entity.

Some governments may find it useful to establish a GIB as a wholly new entity. Independent status provides flexibility, facilitates a focus on targeted objectives, attracts skilled specialists, creates necessary room for innovation and enables authorities to hold the GIB accountable for results. Independence may be secured by issuing a charter for an independent institution, designating a GIB as a non-profit organisation or establishing a subsidiary of an existing institution. Options for structuring a GIB are further discussed below.

The status of a GIB as fully independent or part of an existing administrative structure will affect start-up and operating costs. When studying the feasibility of creating the UK Green Investment Bank, McKinsey estimated start-up costs of GBP 11 million (Vivid Economics and McKinsey & Co, 2011). In its annual reporting, Australia's CEFC is required to disclose detailed operating costs and benchmark these expenses with comparable entities such as the UK Green Investment Bank (CEFC, 2014).

Creating a green investment bank as a wholly new entity

Efforts to establish an entirely new institution may face political resistance if such efforts are viewed as expanding bureaucracy or creating duplicative government services. However, launching a new institution can usefully allow a GIB to establish its own procedures and norms and hire its own staff. For instance, building a focus on LCR investment and preservation of public capital can be easier to achieve in a new institution. In addition, creating a new institution that is independent and free from government interference may be seen as crucial for GIBs that operate on commercial terms. When studying the possibility of creating the UK Green Investment Bank, the Green Investment Bank Commission recognised that rationalising government low-carbon institutions and funds would be helpful in the long-term, but that merging existing programmes would not provide the "game-changing" institution that was needed (UK House of Commons, 2011a). One disadvantage to creating a new entity is that the start-up costs and time investment may be significant.

Converting an existing programme or fund into a green investment bank

A number of GIBs emerged through consolidation of existing renewable energy or green investment programmes. For example, the Connecticut Green Bank was established as a new administrative entity by transferring the net assets and funding sources of the Connecticut Clean Energy Fund. Australia's CEFC incorporated an existing national energy efficiency fund into its structure. Converting an existing entity into a GIB may pose challenges that require new leadership and staff. A government office, for instance, that previously operated a grant programme must willingly accept a change in mission and approach to be successful. An existing government entity may not have existing staff with the financial expertise a GIB requires.

Creating a new GIB as a subsidiary within an existing entity can offer a middle road that provides the benefits of a "blank slate" but also allows the GIB to reduce start-up costs by using the parent entity's accounting and human resources functions. Another option is to create a joint subsidiary. The New Jersey Energy Resilience Bank (ERB), created in July 2014, is jointly administered by the Board of Public Utilities of New Jersey and the state's economic development agency (ERB, 2014). This structure allows the ERB to benefit from the energy-sector knowledge of the utility board and the financing experience of the economic development agency, while still maintaining some operational independence.

Capitalisation and financial sustainability of green investment banks

The first step in capitalising a GIB is to determine the capitalisation amount. For some GIBs this amount was determined based on assessments of market size and funding gaps. Estimates for the amount of investment required and the level of capitalisation needed for the proposed UK Green Investment Bank ranged from GBP 2-7 billion (UK House of Commons, 2011a). For NY Green Bank, analysis by Booz & Company (2013) confirmed that a USD 1 billion capitalisation was consistent with its identified total market size of USD 85 million. For other GIBs and GIB-like entities, the initial capitalisation has been more of a function of available funds (e.g. Swiss Technology Fund, Montgomery County Green Bank).

In most cases, funding for a GIB constitutes a capitalisation, or an infusion of investment capital, that the GIB can use for lending and other return-generating activities. For GIBs that are self-sustaining or profitable, operating expenses may be covered, at least after the initial investment phase, by the returns earned through financing activity. During the start-up period, however, operating losses may be expected and additional funding could be allocated for this purpose. For example, in the regulatory order capitalising NY Green Bank, the Public Service Commission allotted USD 13.2 million for internal and contracted administrative services during the start-up phase (New York Public Service Commission, 2013).

Once a GIB is capitalised, it can act as a revolving loan fund. Like deposits in a commercial bank, the money will be used to make productive loans for which repayment rates assure that the lending bank either makes a profit or breaks even. If the GIB makes equity investments, it can choose whether to hold or try to exit its investments once they have matured. In large-scale infrastructure investments, however, recycling of capital may only be possible over a long time period given the time needed for construction and payback.

Two common sources of GIB capitalisation are funds drawn over time by regulators from the electricity sector, or alternatively, an appropriation of a fixed sum of government funds. However, many other sources of funding have been used or proposed as sources of GIB capitalisation. The appropriateness of a given funding source and the set of possible options will vary significantly based on the local context and political and regulatory environment. Using capital markets to provide capitalisation or subsequent recapitalisations holds significant promise for future GIBs, but to date this approach has only been used by the Hawaiian Green Energy Market Securitization (GEMS) programme. The following sections discuss the diverse sources of initial capitalisation and additional funding for GIBs.

Capitalisation sources

- Government capitalisation: Australia's CEFC receives transfers from the Treasury. Masdar Capital was capitalised directly by the Abu Dhabi government through Mubadala Development Company, a sovereign wealth fund.
- Government grants and programmes: In the United States and other countries with a sub-national structure, state GIBs may be funded by federal grants. For example, New Jersey's Energy Resilience Bank draws upon initial funding from the US Department of Housing and Urban Development, delivered to the state as part of the recovery plan after Hurricane Sandy (Friedrich, 2014).[1]
- Emissions trading schemes: NY Green Bank sourced over USD 50 million from emissions allowance auction proceeds under the Regional Greenhouse Gas Initiative (RGGI) (New York Public Service Commission, 2013). The RGGI also contributes USD 5-10 million annually for the Connecticut Green Bank (Connecticut Green Bank, 2014).
- Utility bill surcharges: The state of New York imposes a system benefits charge on all utility customers. NY Green Bank used a portion of these funds to fund its initial capitalisation. Connecticut levies a USD 0.001 per kWh surcharge on electric ratepayer bills that provides about USD 30 million annually for the Connecticut Green Bank (Connecticut Green Bank, 2014).
- Loans: Most GIBs do not have the ability to borrow. However, the Connecticut Green Bank makes frequent use of majority or wholly-owned special purpose entities which can borrow and to date has raised USD 55 million in non-recourse financing using this technique. The cash flows from residential and commercial-scale loans, leases and power purchase agreements for solar PV are pledged to support these financings (CEFIA, 2013).
- Renewable portfolio standards (RPS) or energy efficiency resource standards (EERS): RPS or renewable electricity standards (RES) are policies that require electricity supply companies to produce a designated percentage of electricity from renewable sources. EERS set targets for energy efficiency savings that utilities must meet and often accompany RPS or are designed to complement renewable energy policies. Non-compliance with these different types of standards can trigger penalties, or "alternative compliance payments", which generate government revenues. NY Green Bank used revenue from RPS and EERS to fund its initial capitalisation (New York Public Service Commission, 2013).

- Bond issuance: The funding source with perhaps the greatest potential is bond issuance, which to date has only been used by Hawaii's GEMS programme to issue USD 150 million in bonds to fully fund its initial capitalisation. Hawaii's bond will be repaid using funds from an existing consumer surcharge on electrical bills (Hawaii Clean Energy Financing Initiative, 2013). Based on traditional "rate reduction bonds" that have typically been used by utilities to finance stranded assets or disaster recovery projects, the AAA-rated GEMS bond issued in 2014 was able to access low-cost capital that is off balance sheet and therefore does not impact the state's budget. GEMS "green bonds" won an award (the 2014 International Financing Review North America Structured Finance Issue Award) for innovative use of the rate reduction bond structure to finance renewable energy infrastructure. The market for "green bonds" – bonds used to fund projects that have environmental or climate benefits – is growing rapidly, with an estimated USD 48 billion in issuance in 2015 (OECD, 2016 forthcoming).

  The UK Green Investment Bank has expressed interest in issuing bonds as a source for continued fundraising. However, under the terms of its establishment it is not permitted to borrow (including by issuing bonds) until 2015-16 and only when the percentage of government debt to GDP begins to fall (UK House of Commons, 2011b). Although the Connecticut Green Bank has not issued bonds to date, it has the authority to issue USD 50 million in bonds backed by a "special capital reserve fund" which effectively guarantees that the state of Connecticut will pay out bond returns and repay bond principle if the Connecticut Green Bank cannot do so.

  Central banks also have the potential to provide GIB funding through their purchase of GIB-issued green bonds. Some central banks have already shown interest in international green bond issuances similar to those that could be made by GIBs. For example, the Brazilian and German central banks invested in the International Finance Corporation's USD 1 billion green bond issuance in 2013 (World Bank, 2014). The governor of the Bank of England has proposed "green quantitative easing" in which the bank would purchase bonds from entities that support low-carbon and climate-resilient (LCR) investment such as the UK Green Investment Bank (Clark and Giles, 2014).

- Carbon tax revenue: The Japanese Ministry of the Environment allocated a portion of carbon tax revenue to fund the creation of its Green Fund (Kaibu, 2013). For the case of the Swiss Technology Fund, a maximum of CHF 25 million per year from CO_2 levy revenue for process and heating fuel is allocated to the fund from 2013 until 2020.

Other potential sources of capitalisation which have not been used to date include the sale of an equity stake and the raising of private capital into GIB managed funds. For example, in June 2015 the UK government proposed the sale of an equity stake into the UK Green Investment Bank to provide additional funds (Box 5.1).

Stable and predictable funding provides private investors with greater certainty

The timing and certainty of capitalisation will have a significant impact on the perceived staying power of a GIB. In the institution's initial years, the level of funding depends significantly on the size of the initial capitalisation and the timing of future re-capitalisations. Some entities receive a large initial capitalisation; for example, Australia's CEFC had AUD 2 billion available in initial capitalisation. Other GIBs start off with more conservative initial capitalisations which require future funding rounds to grow to desired levels. For example, while NY Green Bank expected to have USD 1 billion eventually available for financing, the initial capitalisation in December 2013 accounted for around a fifth of this amount (Klopott, 2013). NY Green Bank's full USD 1 billion capitalisation was finalised in an order issued by the Public Service Commission on 21 January 2016 (New York Public Service Commission, 2016).

Box 5.1. **UK Green Investment Bank may sell an equity stake for further capitalisation**

The announcement in June 2015 that the UK Department for Business, Innovation & Skills would sell a majority equity stake in the UK Green Investment Bank generated significant debate. E3G, the think tank that originally developed the idea of the green investment bank, believes that selling a majority stake will damage investor confidence in the institution and in the government's commitment to developing a low-carbon economy (E3G, 2015). Opponents to the privatisation of the UK Green Investment Bank have raised concerns that its unique dual purpose of achieving profit and green outcomes could be replaced with a primary duty to maximise profits, which could weaken its ability to leverage private investment into more risky low-carbon and climate-resilient infrastructure projects. The government has defended its decision and maintains that regardless of ownership changes, the Green Investment Bank will likely remain both green and profitable, as its green specialisation is probably what will attract investors. It notes that taking on private investors will free the bank from borrowing limitations and compliance with EU state aid regulations, allowing it to access more capital and deploy it across a wider range of green projects (UK Department for Business, Innovation & Skills, 2015a; 2015b).

Amid growing concerns over the preservation of the UK Green Investment Bank's green identity, the UK House of Commons released a report recommending that the bank's privatisation not go ahead unless the bank's green credentials are upheld and stating that the current proposed protections are not "robust enough" (UK House of Commons, 2015).

Sources: E3G (2015), "Green Investment Bank privatisation threatens to undermine UK economic recovery", press release, 24 June, E3G, http://e3g.org/news/media-room/green-investment-bank-privatisation-threatens-uk-economic-recovery; UK Department for Business, Innovation & Skills (2015a), "The future of the Green Investment Bank", speech by The Rt Hon Sajid Javid MP, 25 June, www.gov.uk/government/speeches/the-future-of-the-green-investment-bank; UK Department for Business, Innovation & Skills (2015b), "Future of UK Green Investment Bank PLC: Policy statement, November", BIS/15/630, Crown copyright, London, www.gov.uk/government/uploads/system/uploads/attachment_data/file/477493/BIS-15-630-future-of-the-uk-green-investment-bank.pdf; UK House of Commons (2015), "The future of the Green Investment Bank", UK House of Commons Environmental Audit Committee, Second Report of Session 2015-16, 19 December, The Stationary Office Limited, London, www.publications.parliament.uk/pa/cm201516/cmselect/cmenvaud/536/536.pdf.

Legislation creating a GIB can also define whether or not funding can be revoked or withheld. If a GIB depends on annual budget negotiations through national or sub-national political structures, it could be vulnerable to budget cuts. Australia's CEFC's founding legislation provided for AUD 2 billion in capitalisation per year for five years. In 2014, political opponents sought to defund or shut down the CEFC, but were unable to fully pass the legislation required to impede or halt the CEFC's activity (CEFC, 2013). The Connecticut Green Bank faced a budget challenge in 2013 as the state legislature proposed a significant reallocation of GIB funds (USD 25 million) to the state's general fund (State of Connecticut, 2013). While the bank was able to avoid this fund diversion by offsetting funds transferred with USD 25 million of additional allowance revenues from the RGGI auction proceeds, this example shows the potential risks of funding GIBs through yearly budgets instead of a longer term funding period.

Leadership and staffing

The mission and orientation of GIBs are distinct from those of most existing government agencies. The commercial focus of a GIB requires leadership and staff that reflect its mandate to catalyse new investment and to simultaneously preserve or increase public capital.

Leadership

As GIB leaders interact with experts in the energy market, the financial sector and government, they need to have the ability to navigate the public sector as well as transaction experience. A GIB leader must understand where and how the GIB fits within a broader policy context, including existing renewable energy subsidies or incentives, and needs to anticipate how it could collaborate, or potentially conflict with, existing policies and agencies. With respect to transaction experience, a GIB needs to offer executive salaries and performance incentives that are competitive with the private sector in order to attract well-qualified experts. This can attract criticism; for example, lobby groups and the media have drawn attention to the pay and bonuses for senior UK Green Investment Bank management (Bain, 2015).

Staffing a green investment bank

Due to their focus on mobilising private investment, GIBs need to collaborate, partner and co-invest with a range of private sector actors. GIBs must also be comfortable marketing their role to potential partners. A GIB designed to operate as a wholesale lender which primarily serves large financing institutions will need staff with significant investment experience, while an institution that takes a retail lending approach will require sufficient administrative staff to manage applications for loan or leasing programmes. Technical energy experts and marketing staff may also be required depending on the GIB's strategy and investments. Alternatively, a GIB can take a "lean" staffing approach and rely heavily on contractors or external partners. The Hawaii Green Infrastructure Authority, created in 2014 to manage the GEMS programme, has a team of five which administers the programme with help from private market partners who already have experience and relevant infrastructure in place.

Oversight, reporting and measuring success

Corporate governance and oversight

GIBs are typically overseen by independent governing boards. GIB boards of directors can help insulate GIBs from direct interference from politicians. If a GIB executive is selected by and reports to a board rather than a government official or entity, the institution can be in a better position to weather political changes and operate with a long-term operational view. Board members may be part of or independent of the government. In addition to governance boards, a GIB may also have an advisory board, like that of the Japanese Green Fund, which is more academic in nature.

Reporting and performance metrics

GIBs measure their performance using a range of metrics which generally focus on investment and economic results or climate-related outcomes. Common metrics include total public capital invested, private capital invested in GIB projects, private-to-public leverage ratio, return on capital, energy generated or saved, greenhouse gas (GHG) emission reductions and job creation.

By offering clear performance metrics, GIBs can demonstrate their value and cost-effectiveness. A GIB's performance metrics are typically determined during its creation, either through legislation or regulation. For example, Australia's CEFC is required by law to produce an annual report that includes a set of specific metrics and financial statements (CEFC, 2012), while NY Green Bank's self-created metrics were approved by the state regulatory agency when NY Green Bank was established (NY Green Bank, 2014a). In 2015, the UK Green Investment Bank published a *Green Investment Handbook*, which provides guidance on how it measures green performance, manages risk, conducts due diligence and engages consultants (UK Green Investment Bank, 2015a). To provide transparency for its performance calculations, the UK Green Investment Bank published its "green impact reporting criteria" (UK Green Investment Bank, 2014a). Similarly, the Connecticut Green Bank worked with the state's economic development agency to build a tool to determine the number of jobs created per dollar of investment.

Green banks that focus on profitability or financial sustainability must report detailed financial metrics, as in the examples below[2]:

- In 2013, the Connecticut Green Bank reported that its investments resulted in 27 MW of new renewable generation capacity, avoided 250 000 tonnes of CO_2 lifetime emissions and achieved a 10:1 leverage ratio (i.e. ratio of private investment per USD of public investment) (Connecticut Green Bank, 2013). In 2014, the bank reported a 3:1 leverage ratio (Connecticut Green Bank, 2015).
- The UK Green Investment Bank reported GBP 668 million of new capital commitments in 2014, equating to a 3:1 ratio of private investment per GBP of public investment. The UK Green Investment Bank's investments will generate an estimated 8% rate of return, support 3 500 jobs and reduce GHG emissions by an amount equivalent to taking 1.6 million cars off the road (UK Green Investment Bank, 2014b). In 2015, the UK Green Investment Bank reported a 3:1 leverage ratio for its investments since its inception (UK Green Investment Bank, 2015b).
- Once constructed and in operation, the projects in which Australia's CEFC is investing are estimated to achieve annual emissions abatement of 4.2 million tonnes CO_2-equivalent (tCO_2e), with a net financial return to the CEFC (inclusive of government borrowing costs and operating costs) of approximately AUD 10 million (i.e. emission reductions are achieved at a "cost" of negative AUD 2.40 per tonne[3]) (CEFC, 2014; 2015b). Australia's CEFC is also given specific key performance indicators and associated targets. In 2014, the CEFC achieved a 4.15% return (net of operating costs) on an expected deployed capital of AUD 931 million, exceeding its portfolio benchmark return of 3.14% (CEFC, 2014). As of October 2015, the portfolio of investments in 2015 was projected to generate an annual yield of 6.1% once fully deployed (CEFC, 2015a).[4] As of December 2015, the CEFC's portfolio benchmark return had increased to "at least

the five-year Australian Government bond rate +4 to +5 per cent per annum" (Australian Government, 2015). The CEFC's Board maintains that this target will require the CEFC to "identify and contract out-of-market returns" (CEFC, 2015c) and awaits a review of this policy by the Finance and Environment Ministries in 2016.

Collaboration with other green investment banks

Informal collaboration among GIBs provides the opportunity to share information and lessons learnt. For example, the UK Green Investment Bank hosted a Green Bank Congress in 2013 to bring together institutions and NY Green Bank similarly supported a Green Bank Summit in 2014. The Green Bank Congress now appears to be an annual event, with hosting responsibilities rotating among GIBs. GIBs have also collaborated through staff exchanges. Staff from Australia's CEFC have undertaken temporary assignments at the UK Green Investment Bank, and vice versa, to increase information exchange and share best practices.

Collaboration can also come in the form of formal partnerships. For example, the UK Green Investment Bank and the Green Finance Organisation, which operates the Japanese Green Fund, have entered into a Memorandum of Understanding (MoU) to support information sharing. Masdar and the UK Green Investment Bank also signed an MoU to jointly invest in renewable energy projects in the United Kingdom (Masdar, 2013).

To formalise and expand these collaborations, at the 21st Conference of the Parties to the United Nations Framework Convention on Climate (COP21) in Paris, the UK Green Investment Bank, the Connecticut Green Bank, NY Green Bank, the Green Fund (Japan), the Malaysian Green Technology Corporation and the Clean Energy Finance Corporation (Australia) announced the establishment of a "Green Bank Network". The network "will increase the global impact of green banks by enabling them to collaborate more effectively, share and leverage individual bank experiences, publicize achievements and grow the ranks of green banks worldwide" (UK Green Investment Bank, 2015c). The Natural Resources Defense Council (NRDC) and the Coalition for Green Capital (CGC), two non-governmental organisations with experience in developing green banks, were selected to spearhead the creation of the network and ClimateWorks Foundation provided seed funding (UK Green Investment Bank, 2015c).

Green investment banks as temporary or permanent institutions

As discussed in Chapter 2, GIBs' mandate to avoid "crowding out" private investment requires them to shift into new technologies with less attractive risk-return profiles when their interventions are no longer needed to encourage investment. GIBs can make this shift as part of normal business and fulfilling their mission. For example, NY Green Bank's mission to "transform financing markets" gives it broad authority to invest across a range of developed and emerging target technologies in response to the needs of the market (NY Green Bank, 2014b). However, opinions likely will vary regarding when a GIB should exit a particular market and which deals constitute proof that a GIB's interventions are no longer needed.

Indeed, concerns about public entities "crowding out" private investment are not unique to GIBs, nor are concerns about public entities failing to prove they are mobilising "additional" investment (i.e. investment that would not have occurred without the public entity's involvement). Based on a literature review, the UK Aid Network concluded that relatively little evidence exists for the "financial additionality" of projects using official development assistance to attract private investments. In addition, different entities use distinct methodologies to measure additionality and additionality assessments often lack sufficient detail (UKAN, 2015). A study prepared for the European Parliament's Committee on Development also concluded that existing evidence of the financial additionality of private investment leveraged by public finance was weak (European Parliament, 2014). For example, the study noted that "a systematic review of additionality looking at several MDBs and DFIs, including 17 institutions based in Europe, found that 55% of the projects would have gone ahead without the public finance".

A related issue is whether GIBs are conceived to be permanent rather than temporary institutions that address investment barriers for all targeted sectors and in so doing, eventually ensure their own obsolescence and termination. Technology cost reductions, market evolution and successful efforts by GIBs to catalyse new investment by the private sector will mean that GIB interventions for any particular sector and technology cannot be indefinite. For example, investments in some onshore wind projects may raise additionality issues, depending upon the particular market. However, many technologies that are not yet commercial will likely be needed to meet climate policy objectives. Most GIBs also include several non-commercial technologies in their list of target sectors. In principle, GIBs can shift to different technologies over time and will not run out of investment barriers to address in the near term. This would suggest that GIBs will not be short-lived institutions.

Other factors may limit the life span of GIBs, however. As noted in Chapter 1, GIBs are just one element of the domestic policy framework needed to support the low-carbon transition. Other elements of the framework include fossil fuel subsidy reform; putting a price on carbon; providing well-designed, well-timed, well-targeted and time-limited incentives for renewable energy investment; and setting clear, long-term policy goals. Jurisdictions which make progress in implementing these elements may conclude that creating a new GIB is not warranted or that maintaining an existing GIB is no longer justified. On the other hand, some investment barriers may not be fully addressed by the above-mentioned policies and may require focused interventions by entities like GIBs.

Another challenge to the continuity of some GIBs could be the tension between mandates for profitability and for avoiding "crowding out" private investment. GIBs with mandates to be profitable have a unique challenge: to simultaneously provide sufficient interventions to spur investment in less commercial technologies, achieve targets for financial performance (e.g. through profit-yielding loan repayments they receive) and continue to demonstrate additionality. At some point, for additionality reasons, GIBs with these mandates may be required to focus on technologies for which they may struggle to generate a sufficient return. As GIBs develop a longer track record and more experience in leveraging investment in different technologies, future research could assess the effectiveness of GIBs to date in terms of cost-effectively mobilising private investment, avoiding crowding out private investment, carefully gauging investment risks, effectively targeting and addressing key investment barriers, and successfully demonstrating the viability of LCR infrastructure investment.

Notes

1. The state of New Jersey received USD 1.46 billion in federal funds as part of a Hurricane Sandy recovery package. The state allocated USD 200 million for the creation of the New Jersey Energy Resilience Bank based on an amendment to the second funding allocation (State of New Jersey, 2014).
2. All results are self-reported by GIBs.
3. The CEFC does not claim that the emissions benefit occurs exclusive of other Australian government policy such as the Renewable Energy Target.
4. In 2015, the CEFC had a mid-year change in both its statutory benchmark rate and the method of calculation (see CEFC, 2015c for more information).

References

Australian Government (2015), "Clean Energy Finance Corporation investment mandate direction 2015 (No.2)", Commonwealth of Australia, www.comlaw.gov.au/Details/F2015L02114.

Australian Government (2012), *Clean Energy Finance Corporation Expert Review: Report to Government*, March, Commonwealth of Australia, www.cefcexpertreview.gov.au/content/report/downloads/CEFC_report.pdf.

Bain, S. (2015), "Highly-paid chief brings energy to the first green bank", *The Herald Scotland*, 29 August, www.heraldscotland.com/business/13634321.Highly_paid_chief_brings_energy_to_the_first_green_bank.

Booz & Company (2013), "New York State Green Bank: Business plan development: Final report", Booz & Company, 3 September, www.naseo.org/Data/Sites/1/documents/committees/financing/notes/2013-11-13-Green-Bank-Final-Report.pdf.

CEFC (2015a), "Annual report 2014-2015," Clean Energy Finance Corporation, Sydney, Australia, http://annualreport2015.cleanenergyfinancecorp.com.au.

CEFC (2015b), "CEFC has helped accelerate $3.5b in total investment towards a competitive clean energy economy", press release, 15 July, Clean Energy Finance Corporation, Sydney, Australia, www.cleanenergyfinancecorp.com.au/media/releases-and-announcements/files/cefc-has-helped-accelerate-$35b-in-total-investment-towards-a-competitive-clean-energy-economy.aspx.

CEFC (2015c), "Letter from Jillian Broadbent AO to The Hon Greg Hunt MP Minister for the Environment and Senator the Hon Matthias Cormann Minister for Finance", 1 December, www.cleanenergyfinancecorp.com.au/media/158232/cefc_response_to_investment_mandate_dec_2015.pdf.

CEFC (2014), "Annual report 2013-2014," Clean Energy Finance Corporation, Sydney, Australia, www.cleanenergyfinancecorp.com.au/reports/annual-reports/files/annual-report-2013-14/performance/cefcs-budgeted-outcome-and-key-performance-indicators.aspx.

CEFC (2013), "Submission by the Clean Energy Finance Corporation to the Environment and Communications Reference Committee Inquiry into the government's direct action plan", 12 December, www.cleanenergyfinancecorp.com.au/media/76195/cefc-submission-to-the-environment-and-communications-references-committee-inquiry-into-the-direct-action-plan.pdf.

CEFC (2012), "Clean Energy Finance Corporation Bill 2012", House of Representatives, Parliament of the Commonwealth of Australia, Section 74, Commonwealth of Australia, www.comlaw.gov.au/Details/C2012B00083/Html/Text#_Toc325113802.

CEFC (2011), *Clean Energy Finance Corporation Expert Review: Request for Submissions*, Commonwealth of Australia, www.cefcexpertreview.gov.au/content/consultation/subrequest/CEFCRequest_for_Submissions.pdf.

CEFIA (2013), "CT Solar Lease 2 Program due diligence package," http://resources.ctgreenbank.com/BoardMembers/CGBBoardMeetings/CEFIABoardMeetingMaterials02-15-2013/tabid/702/Default.aspx.

Clark, P. and C. Giles (2014), "Mark Carney boosts green investment hopes", *Financial Times*, 18 March, www.ft.com/cms/s/0/812f3388-aeaf-11e3-8e41-00144feab7de.html#axzz3WR8DlSRq.

Connecticut Green Bank (2015), "Innovating, educating and activating to accelerate clean energy: 2014 annual report", Connecticut Green Bank, Stamford, Connecticut, www.ctgreenbank.com/wp-content/uploads/2015/12/AnnualReport_FINAL_5.4.15-SinglePages.pdf.

Connecticut Green Bank (2014), "Comprehensive annual financial report: Fiscal year ended June 30, 2014", Department of Finance and Administration, Rocky Hill, Connecticut, www.ctgreenbank.com/wp-content/uploads/2015/12/CGB-finalized-financials.pdf.

Connecticut Green Bank (2013), "Connecticut's Green Bank, energizing clean energy finance: 2013 annual report", Connecticut Green Bank, Stamford, Connecticut.

E3G (2015), "Green Investment Bank privatisation threatens to undermine UK economic recovery", press release, 24 June, E3G, http://e3g.org/news/media-room/green-investment-bank-privatisation-threatens-uk-economic-recovery.

ECRI (2014), "ECRI statement on a possible 'green bank'", Environmental Council of Rhode Island, www.ecori.org/s/ECRIGreenBank.pdf.

ERB (2014), "New Jersey Energy Resilience Bank Grant and Loan Financing Program Guide", New Jersey Energy Resilience Bank, www.state.nj.us/bpu/pdf/erb/FINAL%20DRAFT%20-%20ERB%20Program%20Guide%208%2020%2014.pdf.

European Parliament (2014), *Financing for Development Post-2015: Improving the Contribution of Private Finance,* the European Parliament's Directorate-General for External Policies, April, European Union, www.europarl.europa.eu/RegData/etudes/etudes/join/2014/433848/EXPO-DEVE_ET%282014)433848_EN.pdf.

Friedrich, K. (2014), "Clean energy and climate resilience join forces", Clean Energy Finance Forum, 7 March, http://cleanenergyfinanceforum.com/2014/03/07/clean-energy-and-climate-resilience-join-forces/?utm_source=Clean+Energy+Finance+Forum+3%2F12%2F2014+-+Issue+%2349&utm_campaign=3%2F12%2F2014&utm_medium=email.

Hawaii Clean Energy Initiative (2013), "State loans urged to help residents install solar gear", 19 February, www.hawaiicleanenergyinitiative.org/state-loans-urged-to-help-residents-install-solar-gear (accessed 30 March 2014).

Kaibu, A. (2013), "Green fund launched to accelerate low-carbon investments", *Japan Environment Quarterly*, Vol. 2, www.env.go.jp/en/focus/jeq/issue/vol02/topics.html#c7.

Klopott, F. (2013), "Cuomo starts $1 billion New York Green Bank for energy lending", Bloomberg, 10 September, www.bloomberg.com/news/2013-09-10/cuomo-starts-1-billion-new-york-green-bank-for-energy-lending.html.

Liberal Party of Australia (2013), "Letters the Hon Tony Abbott MHR Leader of the Opposition to Dr Ian Watt AO Secretary Department of Prime Minister & Cabinet and to Ms Jilian Broadbent AO Chair Clean Energy Finance Corporation", 5 August, http://static.liberal.org.au.s3.amazonaws.com/13-08-05%20Signed%20Carbon%20Tax%20Letters%20-%20TA.pdf.

Masdar (2013), "Masdar and the UK Green Investment Bank form new project investment alliance", press release, Masdar Institute, 1 May, http://masdar.ae/research/detail/masdar-and-uk-green-investment-bank-form-new-project-investment-alliance.

New York Public Service Commission (2016), "Order authorizing the clean energy fund framework", 21 January, http://documents.dps.ny.gov/public/Common/ViewDoc.aspx?DocRefId={C766B4CA-D6F7-4A14-8869-5FFE4412BEA3}.

New York Public Service Commission (2013), "Petition of the New York State Energy Research and Development Authority to provide initial capitalization for the New York Green Bank", Case 13-M, State of New York Public Service Commission, 9 September, http://documents.dps.ny.gov/public/Common/ViewDoc.aspx?DocRefId=%7BC3FE36DC-5044-4021-8818-6538AAB549B8%7D.

New York State (2013), "Governor Cuomo launches New York Green Bank initiative to transform the state's clean energy economy", press release, New York State, 10 September, www3.dps.ny.gov/pscweb/WebFileRoom.nsf/Web/17D6A0944B66384C85257BE20063731E/$File/Gov%209-10-13.pdf?OpenElement.

NY Green Bank (2014a), *Metrics, Reporting & Evaluation Plan*, New York Green Bank, New York Public Service Commission Case 13-M-0412, 19 June, http://documents.dps.ny.gov/public/Common/ViewDoc.aspx?DocRefId=%7B6F5D6757-CACA-4A57-B1F2-0DCA5F4C5946%7D.

NY Green Bank (2014b), *Business Plan*, Case 13-M-0412,http://documents.dps.ny.gov/public/Common/ViewDoc.aspx?DocRefId=%7B3BCF6C87-33FB-49FA-B264-8669055DD6E5%7D.

OECD (2016 forthcoming), *Mobilising the Debt Capital Markets for a Low-Carbon Transition*, OECD Publishing, Paris, forthcoming.

Proft, K. (2015), "Gov. Raimondo's green bank idea draws some concern", *eco RI news*, 30 January, www.ecori.org/government/2015/1/30/gov-raimondos-green-bank-idea-draws-some-concern.

State of Connecticut (2013), House Bill No. 6704, Public Act No. 13-184, Hartford, Connecticut, www.cga.ct.gov/2013/act/pa/2013PA-00184-R00HB-06704-PA.htm.

State of New Jersey (2014), "Christie administration rolls out plan for second round of federal Sandy recovery funds", press release, State of New Jersey, 3 February, http://nj.gov/governor/news/news/552014/approved/20140203a.html.

UKAN (2015), "Leveraging aid: A literature review on the additionality of using ODA to leverage private investment", report prepared by the UK Aid Network, www.ukan.org.uk/wordpress/wp-content/uploads/2015/03/UKAN-Leveraging-Aid-Literature-Review-03.15.pdf.

UK Department for Business, Innovation & Skills (2015a), "The future of the Green Investment Bank", speech by The Rt Hon Sajid Javid MP, 25 June, www.gov.uk/government/speeches/the-future-of-the-green-investment-bank.

UK Department for Business, Innovation & Skills (2015b), "Future of UK Green Investment Bank Plc: Policy statement, November", BIS/15/630, Crown copyright, London, www.gov.uk/government/uploads/system/uploads/attachment_data/file/477493/BIS-15-630-future-of-the-uk-green-investment-bank.pdf.

UK Green Investment Bank (2015a), *Green Investment Handbook*, United Kingdom Green Investment Bank, Edinburgh, www.greeninvestmentbank.com/green-impact/green-investment-handbook.

UK Green Investment Bank (2015b), *Annual Report 2014-15*, UK Green Investment Bank Plc, Edinburgh, www.greeninvestmentbank.com/about-us/2015-annual-review.

UK Green Investment Bank (2015c), "First global Green Bank Network will speed shift to clean energy", News, 9 December, www.greeninvestmentbank.com/news-and-insight/2015/first-global-green-bank-network-will-speed-shift-to-clean-energy.

UK Green Investment Bank (2014a), "Green impact reporting criteria", Green Investment Bank, Edinburgh, June, www.greeninvestmentbank.com/media/25370/green-impact-reporting-criteria_final.pdf.

UK Green Investment Bank (2014b), *Annual Report 2014*, UK Green Investment Bank, Edinburgh, www.greeninvestmentbank.com/media/25360/ar14-web-version-v2-final.pdf.

UK House of Commons (2015), "The future of the Green Investment Bank", UK House of Commons Environmental Audit Committee, Second Report of Session 2015-16, 19 December, The Stationary Office Limited, London, www.publications.parliament.uk/pa/cm201516/cmselect/cmenvaud/536/536.pdf.

UK House of Commons (2011a), "The Green Investment Bank", Second Report of Session 2010-11, Volume I: Report, together with formal minutes, oral and written evidence, 11 March, House of Commons, The Stationary Office, London, www.publications.parliament.uk/pa/cm201011/cmselect/cmenvaud/505/505.pdf.

UK House of Commons (2011b), *Budget 2011*, Return to an order of the House of Commons by the Chancellor of the Exchequer when opening the Budget, HC 836, The

Stationary Office, London, www.gov.uk/government/uploads/system/uploads/attachment_data/file/247483/0836.pdf.

Vivid Economics and McKinsey & Co (2011), "The economics of the Green Investment Bank: Costs and benefits, rationale and value for money", report prepared for the Department for Business, Innovation & Skills, October, www.gov.uk/government/uploads/system/uploads/attachment_data/file/31741/12-554-economics-of-the-green-investment-bank.pdf.

World Bank (2014), "Growing the green bond market to finance a cleaner, resilient world", News, World Bank, 4 March, www.worldbank.org/en/news/feature/2014/03/04/growing-green-bonds-market-climate-resilience.

Annex 5.A1.
History of formation of green investment banks

This annex provides background information and history on the formation of selected green investment banks (GIBs).

Clean Energy Finance Corporation, Australia

In July 2011, the Australian Prime Minister, Deputy Prime Minister, Treasurer and Minister for Climate Change and Energy Efficiency announced the government's renewable energy plan. The Clean Energy Finance Corporation (CEFC) was part of the government's "Securing a Clean Energy Future" package, a comprehensive plan for carbon pricing, reducing pollution in the land sector and promoting innovation in renewable energy (Australian Government, 2011a).

In October 2011, the government appointed a team of experts to advise on the design of the CEFC and provide recommendations regarding implementation, investment mandates, risk management and governance (Australian Government, 2011b). The expert panel consulted widely with industry and stakeholders and delivered a report to the Australian government in March 2012. The Expert Review also set out a detailed timeline of key tasks to undertake leading up the implementation of the CEFC (Australian Government, 2012a). In July 2012, the Clean Energy Finance Corporation Act 2012 set out the terms for the CEFC's establishment and operation (Australian Government, 2012b). The CECF has a two-pronged funding approach: operational funding is received through parliamentary appropriations while investment funding is set aside in a dedicated Treasury fund, with funds made available when investments are identified. The CEFC's first full-year investments totalled AUD 931 million (CEFC, 2014).

NY Green Bank, United States

Around 80% of the USD 1.4 billion per year spent by New York state entities to promote renewable energy and energy efficiency was in the form of one-time subsidies or grants as of 2013 (New York Public Service Commission, 2013a). Proponents of a GIB envisioned it as a tool to transition away from an unsustainable subsidy-based model to a private market approach that would use limited public capital. In the January 2013 State of the State Address, the Governor of New York called for the establishment of a USD 1 billion New York Green Bank to mobilise private capital to finance the transition to a more cost-effective, resilient and renewable energy system (State of New York, 2013).

The New York State Energy Research and Development Authority (NYSERDA) retained the consulting firm Booz & Company to undertake a market assessment of existing barriers to renewable energy finance, identify financial products to respond to the market and provide recommendations on the organisational structure of the future NY Green Bank. Booz & Company conducted nearly 90 interviews with financial institutions, renewable energy providers, energy service companies, utilities and end users. Based on the identified financing barriers, specific NY Green Bank offerings were proposed to address these specific market gaps. A detailed market sizing by technology identified a total market size of approximately USD 85 billion for renewable energy projects in New York (Booz & Company, 2013). Quantitative modelling also provided information on the expected return on investment and amount of private capital that can be mobilised based on different product offerings. Booz & Company (2013) proposed a timeline for development with key activities, milestones and performance indicators for the establishment and implementation phases.

Supported by this data and analysis, NYSERDA requested in September 2013 the reallocation of USD 165 million in uncommitted funds from energy efficiency and renewable energy portfolio standards and systems benefits charges to fund the initial capitalisation of NY Green Bank (New York Public Service Commission, 2013a). In the autumn of 2013, NYSERDA also engaged stakeholders such as businesses, financial institutions, environmental actors and public sector institutions, among others, in an open public commenting period. The collected comments are available publicly online (New York Department of Public Service, n.d.).

In December 2013, the New York Public Service Commission granted NYSERDA's request and provided NY Green Bank with USD 165.6 million to begin operations (New York Public Service Commission, 2013b). NY Green Bank officially opened for business in February 2014 and has since prepared a detailed request for proposal to the market, submitted an organisational plan (NY Green Bank, 2014a), developed a strategic business plan (NY Green Bank, 2014b), and created specific and detailed performance metrics (NY Green Bank, 2014c). The metrics were subject to a public review and input process. Following several additional open public comment periods, NY Green Bank's full capitalisation of USD 1 billion was finalised in an order issued from the Public Service Commission on 21 January 2016 authorising the creation of a 10-year USD 5 billion clean energy fund (New York Public Service Commission, 2016). In addition to providing the USD 782 million missing for NY Green Bank's targeted capitalisation of USD 1 billion, the fund supports research and innovation, market development and an existing programme (NY-Sun) to support the development of the solar PV market in the state of New York (New York State, 2016).

UK Green Investment Bank

In 2009, Climate Change Capital and E3G published a series of papers which examined how the UK government could mobilise private investment for the low-carbon transition. One of the recommendations was the establishment of a Green Infrastructure Bank (Holmes and Mabey, 2009). Various other organisations including Friends of the Earth, Policy Exchange and the Aldersgate Group also discussed the proposal and published related papers. In February 2010, a working group, the Green Investment Bank Commission, was created. After a vigorous grassroots campaign and the publication of more papers advocating for the establishment of a GIB, in July 2010 the commission published its own report, "Unlocking investment to deliver Britain's low carbon future", recommending that a GIB be established (Green Investment Bank Commission, 2010).

In August 2010, the UK government set up a formal Green Investment Bank Working Group. A month later, Ernst & Young produced a detailed report on the size of the green investment bank and recommended that GBP 4-6 billion would be needed over four years (Ernst & Young, 2010). In October 2010, GBP 1 billion was allocated under the UK Comprehensive Spending Review. In March 2011, an additional GBP 2 billion was allocated bringing the UK Green Investment Bank's initial capitalisation to GBP 3 billion (Holmes, 2013). In May 2012, the creation of the GIB was included in the Enterprise and Regulatory Reform Bill, which was published in June 2012. A few months later, the government published its "Update on Green Investment Bank", which outlined the UK Green Investment Bank's mission, business model and strategic priorities.

From 2011 to the announcement of the UK Green Investment Bank's CEO and Board in September 2012, the UK Green Investment Bank operated with limited staff as a shadow institution called UK Green Investments. During this time, it made indirect investments totaling GBP 180 million. In July 2012, Lord Smith of Kelvin was appointed as the Green Investment Bank's Chairman; in September 2012, Shaun Kingsbury was appointed as CEO. The Enterprise and Regulatory Reform Act, which formalised the establishment of the UK Green Investment Bank, entered into force in April 2013.

References

Australian Government (2012a), *Clean Energy Finance Corporation Expert Review: Report to Government*, March, Commonwealth of Australia, www.cefcexpertreview.gov.au/content/report/downloads/CEFC_report.pdf.

Australian Government (2012b), "Clean Energy Finance Corporation Bill 2012, A bill for an act to establish the Clean Energy Finance Corporation, and for related purposes", No. 104, Commonwealth of Australia, www.comlaw.gov.au/Details/C2012B00083.

Australian Government (2011a), "Securing a clean energy future for Australia", joint media release with Prime Minister and Minister for Climate Change and Energy Efficiency, Canberra, http://ministers.treasury.gov.au/DisplayDocs.aspx?doc=pressreleases/2011/085.htm&pageID=003&min=wms&Year=&DocType=0.

Australian Government (2011b), "Experts to advise on Clean Energy Finance Corporation", joint media release with Prime Minister and Minister for Climate Change and Energy Efficiency, Canberra, 12 October, http://ministers.treasury.gov.au/DisplayDocs.aspx?doc=pressreleases/2011/121.htm&pageID=003&min=wms&Year=&DocType=.

Booz & Company (2013), "New York State Green Bank: Business plan development: Final report", Booz & Company, 3 September, www.naseo.org/Data/Sites/1/documents/committees/financing/notes/2013-11-13-Green-Bank-Final-Report.pdf.

CEFC (2014), "Annual report 2013-2014", Clean Energy Finance Corporation, Sydney, Australia, www.cleanenergyfinancecorp.com.au/reports/annual-reports/files/annual-report-2013-14.aspx.

Ernst & Young (2010), "Capitalising the green investment bank: Key issues and next steps", Ernst & Young LLP, London, http://e3g.org/docs/capitalisingthegreeninvestmentbank.pdf.

Green Investment Bank Commission (2010), "Unlocking investment to deliver Britain's low carbon future", Green Investment Bank Commission, London, www.e3g.org/docs/Unlocking_investment_to_deliver_Britains_low_carbon_future_-_Green_Investment_Bank_Commission_Report_June_2010.pdf.

Holmes, I. (2013), "Green Investment Bank: The history", E3G, 25 April, www.e3g.org/library/green-investment-bnak-the-history.

Holmes, I. and N. Mabey (2009), "Accelerating green infrastructure financing: Outline proposals for UK green bonds and infrastructure", Briefing Note, March, Climate Change Capital and E3G, www.e3g.org/docs/Accelerating_Green_Infrastructure_Financing.pdf.

New York Department of Public Service (n.d.), "Petition of New York State Energy Research and Development Authority to provide initial capitalization for the New York Green Bank", Matter Number 13-01875, New York State Energy Research and Development Authority, http://documents.dps.ny.gov/public/MatterManagement/CaseMaster.aspx?MatterCaseNo=13-m-0412&submit=Search+by+Case+Number.

New York Public Service Commission (2016), "Order authorizing the clean energy fund framework", 21 January, http://documents.dps.ny.gov/public/Common/ViewDoc.aspx?DocRefId={C766B4CA-D6F7-4A14-8869-5FFE4412BEA3}.

New York Public Service Commission (2013a), "Petition of the New York State Energy Research and Development Authority to provide initial capitalization for the New York Green Bank", Case 13-M, State of New York Public Service Commission, 9 September, http://documents.dps.ny.gov/public/Common/ViewDoc.aspx?DocRefId=%7BC3FE36DC-5044-4021-8818-6538AAB549B8%7D.

New York Public Service Commission (2013b), "Order Establishing New York Green Bank and providing initial capitalization", Public Service Commission, Case 13-M-0412, Petition of New York State Energy Research and Development Authority to provide initial capitalization for the New York Green Bank, Albany, New York, 19 December, http://documents.dps.ny.gov/public/Common/ViewDoc.aspx?DocRefId=%7BBD3AAFB0-FAA2-4DA6-B56B-0FF22EE34EDF%7D.

New York State (2016), "Cuomo launches $5 billion Clean Energy Fund to grow New York's clean energy economy", New Release, 21 January, www.governor.ny.gov/news/governor-cuomo-launches-5-billion-clean-energy-fund-grow-new-york-s-clean-energy-economy.

NY Green Bank (2014a), "NY Green Bank organization plan", Filing for the Public Service Commission, Case 13-M-0412, 18 February, http://documents.dps.ny.gov/public/Common/ViewDoc.aspx?DocRefId=%7B643D622C-6E1F-4BD1-A489-DD146F26E8A4%7D.

NY Green Bank (2014b), Business Plan, Case 13-M-0412, http://documents.dps.ny.gov/public/Common/ViewDoc.aspx?DocRefId=%7B3BCF6C87-33FB-49FA-B264-8669055DD6E5%7D.

NY Green Bank (2014c), "Metrics, reporting & evaluation plan, New York Green Bank", New York Public Service Commission Case 13-M-0412, 19 June, http://documents.dps.ny.gov/public/Common/ViewDoc.aspx?DocRefId=%7B6F5D6757-CACA-4A57-B1F2-0DCA5F4C5946%7D.

State of New York (2013), "Governor Cuomo launches New York Green Bank initiative to transform the state's clean energy economy", press release, New York State, 10 September, www3.dps.ny.gov/pscweb/WebFileRoom.nsf/Web/17D6A0944B66384C85257BE20063731E/$File/Gov%209-10-13.pdf?OpenElement.

Glossary[1]

Bankable: Projects that have sufficient collateral, probability of success and predictability of future cash flows to be acceptable to prospective financiers.

Capital recycling: Providing refinancing once a project is at the operational stage so that early-stage investors have an "exit strategy", allowing them to free up capital to invest in new projects, i.e. "recycle" their capital.

Corporate financing: The act or process through which a corporation raises or obtains capital.

Credit enhancement: Reducing the credit or default risk of a debt, thereby improving its credit-worthiness and increasing the overall credit rating.

"Crowding in": Crowding in occurs when public investment induces greater private investment than would have occurred otherwise.

"Crowding out": Crowding out occurs when a public intervention directly displaces the efforts of the private sector by undertaking projects the private sector would have otherwise carried out. Crowding out can also occur indirectly if governments use distortionary taxes to fund public investment, or in situations where demand for government borrowing results in increased interest rates and can make borrowing too costly for private investors.

Crowdsourcing: The process of obtaining ideas, content or funding, usually online, from a large group of people. In the context of this publication, crowdsourcing refers to attracting small unaccredited investors to provide funding for renewable energy projects.

ESCO: An energy service company (ESCO) can offer a broad range of energy services to end-users, including the design and implementation of energy-savings projects, retrofitting, energy conservation, energy infrastructure outsourcing, power generation, energy supply and risk management. What characterises these companies from others is that they can arrange financing where their remuneration can be directly linked to the energy savings achieved.

Green investment bank: Broadly defined as a publicly capitalised entity established specifically to facilitate and attract private investment in domestic low-carbon and climate-resilient infrastructure through different activities and interventions.

Institutional investor: Institutional investors are usually synonymous with "intermediary investors", that is, institutions that manage and invest other people's money. The term institutional investor can be used to describe insurance companies,

1. *Disclaimer:* Explanations on the terms are very condensed and may not be complete. They are not considered to necessarily reflect the official position of the OECD.

investment funds, pension funds, public pension reserve funds (social security systems), foundations and endowments, among others.

Investment bank: An investment bank traditionally facilitates transactions of all types in the wholesale financial markets (transactions conducted by corporations, businesses, institutional investors and high net worth individuals) including mergers and acquisitions (the purchase and sale of businesses and their assets), capital raising or "underwriting" (of equity, debt, etc.) on behalf of corporations or their shareholders. They may provide ancillary services, such as market making; trading of derivatives, securities and other financial instruments; investing and lending; asset management; and fixed income instruments, currencies and commodities (FICC) services. This excludes retail brokerage, retail lending or any other practice that centres on "unaccredited investors".

Liquidity Risk: A financial risk stemming from the lack of marketability of an asset, commodity or security that cannot be converted swiftly enough to preclude an inordinate loss.

Mezzanine financing: Mezzanine financing is senior to common shares (equity) (i.e. mezzanine investors receive returns from the investment before equity holders), but junior to secured debt or senior debt. Mezzanine financing normally includes subordinated (i.e. junior) debt or preferred equity (i.e. equity shares that provide dividends before common stock dividends are paid out) and is usually more expensive than senior debt. It can be used as the stage of financing that follows venture capital.

On-bill finance: On-bill finance allows utility consumers to invest in energy efficiency improvements and repay the funds through additional charges on their utility bill. Under this approach, a third party (such as an energy provider) provides upfront funding for energy efficiency improvements to an investor (e.g. a tenant in a residential or commercial building). The beneficiary pays back the loan via its existing energy bill. In many cases, repayments are structured in such a way that the monthly energy savings achieved through the investment equal or outweigh the loan repayments. If structured properly, an on-bill finance programme can substantially reduce the cost of and improve access to financing.

Origination: Loan origination generally includes all the steps from accepting a loan application up to the disbursal of funds (or denial of the loan application).

Public financial institution (PFI): A publicly created or mandated financial institution created in many cases to correct for the lack of market-based finance through the provision of missing financial services.

Retrofit: An energy efficiency retrofit is an improvement made to an existing structure which improves the overall energy efficiency of a building or home.

Risk mitigant: Risk mitigants include a range of targeted interventions generally aimed at reducing, reassigning or reapportioning different investment risks using mechanisms such as guarantees and insurance products, public stakes and other forms of credit enhancement. By providing coverage for risks which are new and are not currently covered by financial actors, or are simply too costly for investors, risk-mitigating tools increase the attractiveness and acceptability of sustainable energy projects for institutional investors that are particularly risk-averse (e.g. pension funds).

Risk profile: An assessment of the degree to which an investor is prepared to accept losses at the expense of potential gain.

Securitisation: Securitisation is the process of transforming illiquid financial assets into tradable products.

Tax lien: A legal claim by a government entity against a property if tax debts are unpaid. Tax liens are a last resort to force an individual or business to pay back taxes. Tax liens take precedence over all other liens on a property and (in case of liquidation) must be satisfied first.

Transaction enabler: A process or technique which facilitates investment by reducing the associated transaction costs.

Underwriting: In the case of loans, underwriting is the process by which a lender decides whether a potential creditor is creditworthy and should receive a loan. For securities issuances, underwriting is the procedure by which an underwriter, such as an investment bank, brings a new security issue to the investing public in an offering. In such a case, the underwriter will guarantee a certain price for a certain number of securities to the party that is issuing the security (in exchange for a fee). Thus, the issuer is secure that they will raise a certain minimum from the issue, while the underwriter bears the risk of the issue.

ORGANISATION FOR ECONOMIC CO-OPERATION AND DEVELOPMENT

The OECD is a unique forum where governments work together to address the economic, social and environmental challenges of globalisation. The OECD is also at the forefront of efforts to understand and to help governments respond to new developments and concerns, such as corporate governance, the information economy and the challenges of an ageing population. The Organisation provides a setting where governments can compare policy experiences, seek answers to common problems, identify good practice and work to co-ordinate domestic and international policies.

The OECD member countries are: Australia, Austria, Belgium, Canada, Chile, the Czech Republic, Denmark, Estonia, Finland, France, Germany, Greece, Hungary, Iceland, Ireland, Israel, Italy, Japan, Korea, Luxembourg, Mexico, the Netherlands, New Zealand, Norway, Poland, Portugal, the Slovak Republic, Slovenia, Spain, Sweden, Switzerland, Turkey, the United Kingdom and the United States. The European Union takes part in the work of the OECD.

OECD Publishing disseminates widely the results of the Organisation's statistics gathering and research on economic, social and environmental issues, as well as the conventions, guidelines and standards agreed by its members.

OECD PUBLISHING, 2, rue André-Pascal, 75775 PARIS CEDEX 16
(97 2015 35 1 P) ISBN 978-92-64-24511-2 – 2016

www.ingramcontent.com/pod-product-compliance
Lightning Source LLC
LaVergne TN
LVHW081409110826
845149LV00010B/1684
9789264245112